A POSTERIORI ERROR ESTIMATION IN FINITE ELEMENT ANALYSIS

MARK AINSWORTH and J. TINSLEY ODEN

A Wiley-Interscience Publication
JOHN WILEY & SONS, INC.
New York • Chichester • Weinheim • Brisbane • Singapore • Toronto

This text is printed on acid-free paper. ♾

Library of Congress Cataloging in Publication Data is available.

Ainsworth, M., 1965–
A posteriori error estimation in finite analysis / M. Ainsworth and J. T. Oden.
p. cm. — (Pure and applied mathematics)
"A Wiley-Interscience publication."
Includes bibliographical references and index.
ISBN 0-471-29411-X (cloth : alk. paper)
1. Finite element method. 2. Error analysis (Mathematics) I. Oden, J. Tinsley (John Tinsley), 1936– II. Title. III. Pure and applied mathematics (John Wiley & Sons : Unnumbered)
TA 347.F5 A335 2000
620'.001'51535—dc21 00-039915

Printed in the United States of America.

10 9 8 7 6 5 4 3 2 1

To Irene and Sara

Contents

Preface

The combination of powerful modern computers with effective numerical procedures, particularly finite element methods, have transformed what were once purely qualitative theories of mechanics and physics into effective bases for simulation of physical phenomena important in countless engineering applications and scientific predictions. A major problem with such simulations is this: No matter how sophisticated and appropriate the mathematical models of an event, all computational results obtained using them involve numerical error. Discretization error can be large, pervasive, unpredictable by classical heuristic means, and can invalidate numerical predictions. For these reasons, a mathematical theory for estimating and quantifying discretization error is of paramount importance to the computational sciences. Equally important, knowledge of such errors, their magnitude, and their distribution provides a basis for adaptive control of the numerical process, the meshing, the choice of algorithms, and, therefore, the efficiency and even the feasibility of the computation. Understandably, the theory of *a posteriori* error estimation has become an important area of research and has found application in an increasing number of commercial software products and scientific programs.

This book is intended to provide an introduction to the subject of *a posteriori* error estimation of finite element approximations of partial differential equations. It is the outgrowth of over fifteen years of work by the authors on this subject and around a decade of joint work. The book attempts to cover the mathematical underpinnings of many of the most effective methods for error estimation that are available today. It is our aim here to provide a systematic treatment to the theory of error estimation accessible to contemporary engineers and applied scientists who wish to learn not only the mathematical foundations of error estimation, but also the details of their implementation on boundary value problems of continuum mechanics and physics. To make the ideas clear and understandable, we have focused primarily on model scalar elliptic problems on two-dimensional domains. However, we ultimately present significant generalizations to unsymmetric, indefinite problems and to representative nonlinear problems, including the Navier–Stokes equations.

This book is not a treatise; it is instead an introduction of limited scope that does not address a number of subjects now in early stages of development. These include time-dependent problems and many other significant nonlinear problems in mechanics and physics. However, we feel the methods and techniques covered here will provide the basis for work on many other problem areas.

This book started out as an extension of an earlier monograph the authors wrote on this subject that appeared in *Computational Mechanics Advances: Computer Methods in Applied Mechanics and Engineering,* 1997. We are grateful to Elsevier Scientific Publishers for giving permission to reproduce some sections and figures from that earlier article in the present work. The earlier work has been completely revised. We have also added a great deal of new material, including detailed analyzes of hierarchical bases for error estimation, new work on recovery methods, and, particularly, a new account of local and pollution errors, including systematic techniques for estimating so-called errors in local quantities of interest. This theory leads to upper and lower bounds on error estimates of local quantities of interest, such as average gradients, stresses, and solution values, on interior patches of the domain. The book is organized as follows: We begin with an introductory chapter in which our basic notations and assumptions, along with relevant finite element concepts, are presented. We then present successively five distinct techniques of error estimation that have been developed since the late 1970s. We have included chapters on explicit error estimators, recovery methods, implicit error estimators, and the use of hierarchical bases for error estimation, as well as an entire chapter on the equilibrated residual method. We also present techniques for computing estimates in L_2 and L_∞ and in other Sobolev norms. Chapter 7 is devoted to techniques for judging the performance of the various estimators. There we provide a rather detailed framework for the study of the performance of various types of error estimators along with illustrative examples. Chapter 8 provides an account of the estimation of errors in quantities of interest, along with a discussion of local and pollution errors. There we also include numerical examples to demonstrate the performance of local *a posteriori* errors. Chapter 9 is devoted to a number of extensions and applications of the theory. These include (a) nonself-adjoint and indefinite problems such as are embodied in the Stokes problem and Oseen's problem and (b) extensions to linear elliptic systems such as those found in elasticity. We give examples of extensions to nonlinear problems such as the problem of flow characterized by the steady-state Navier–Stokes equations for incompressible fluids.

M. AINSWORTH
Glasgow, Scotland

J.T. ODEN
Austin, Texas

Acknowledgments

It is a pleasure to acknowledge the generous help of a number of colleagues in the preparation of this work. We are indebted to Ivo Babuška for his generous advice and comments on our work over the last decade. We have benefitted from many discussions with Leszek Demkowicz and Theofanis Strouboulis over the years. We are grateful to several scholars who have assisted by reading various portions of the manuscript and offering many valuable comments and suggestions. In particular, we would like to thank Don Kelly, Ralf Kornhuber, Trond Kvamsdal, Halgeir Melbøe, Ricardo Nochetto, and Hans-Goerg Roos, but especially Willy Dörfler and Serge Prudhomme, who read the entire manuscript. Pergamon Press were kind enough to give their permission to reprint some of the figures used in Chapter 8.

M.A. and J.T.O.

A POSTERIORI ERROR ESTIMATION IN FINITE ELEMENT ANALYSIS

1

Introduction

1.1 A POSTERIORI ERROR ESTIMATION: THE SETTING

The presence of numerical error in calculations has been a principal source of concern since the beginning of computer simulations of physical phenomena. Numerical error is intrinsic in such simulations: The discretization process of transforming a continuum model of physical phenomena into one manageable by digital computers cannot capture all of the information embodied in models characterized by partial differential equations or integral equations. What is the approximation error in such simulations? How can the error be measured, controlled, and effectively minimized? These questions have confronted computational mechanicians, practitioners, and theorists alike since the earliest applications of numerical methods to problems in engineering and science.

Concrete advances toward the resolution of such questions have been made in the form of theories and methods of *a posteriori* error estimation, whereby the computed solution itself is used to assess the accuracy. The remarkable success of some *a posteriori* error estimators has opened a new chapter in computational mathematics and mechanics that could revolutionize the subjects. By effectively estimating error, the possibility of controlling the entire computational process through new adaptive algorithms emerges. Fresh criteria for judging the performance of algorithms become apparent. Most importantly, the analyst can use *a posteriori* error estimates as an independent measure of the quality of the simulation under study.

The present work is intended to provide an introduction to the subject of *a posteriori* error estimation for finite element approximations of boundary value problems in mechanics and physics. The treatment is by no means exhaustive, focusing primarily on elliptic partial differential equations and on the chief methods currently available. However, extensions to unsymmetrical systems of partial differential equations, nonlinear problems, and indefinite problems are included. Our aim is to present a coherent summary of *a posteriori* error estimation methods.

1.2 STATUS AND SCOPE

The *a priori* estimation of errors in numerical methods has long been an enterprise of numerical analysts. Such estimates give information on the convergence and stability of various solvers and can give rough information on the asymptotic behavior of errors in calculations as mesh parameters are appropriately varied. Traditionally, the practitioner using numerical simulations, while aware that errors exist, is rarely concerned with quantifying them. The quality of a simulation is generally assessed by physical or heuristic arguments based on the experience and judgment of the analyst. Frequently such arguments are later proved to be flawed.

Some of the earliest *a posteriori* error estimates used in computational mechanics were in the solution of ordinary differential equations. These are typified by predictor–corrector algorithms in which the difference in solutions obtained by schemes with different orders of truncation error is used as rough estimates of the error. This estimate can in turn be used to adjust the time step. It is notable that the original *a posteriori* error estimation schemes for elliptic problems had many features that resemble those for ordinary differential equations.

Modern interest in *a posteriori* error estimation for finite element methods for two point elliptic boundary value problems began with the pioneering work of Babuška and Rheinboldt [22]. *A posteriori* error estimation techniques were developed that delivered numbers η_K approximating the error in energy or an energy norm on each finite element K. These formed the basis of adaptive meshing procedures designed to control and minimize the error. During the period 1978–1983, a number of results for explicit error estimation techniques were obtained: We mention references [20] and [21] as representative of the work.

The use of complementary energy formulations for obtaining *a posteriori* error estimates was put forward by de Veubeke [52]. However, the method failed to gain popularity being based on a global computation. The idea of solving element by element complementary problems together with the important concept of constructing *equilibrated boundary data* to obtain error estimates was advanced by Ladevèze and Leguillon [77]. Related ideas are found in the work of Kelly [73] and Stein [103].

In 1984, an important conference on adaptive refinement and error estimation was held in Lisbon [32]. At that meeting, several new developments in *a posteriori* error estimation were presented, including the *element residual method.* The method was described by Demkowicz *et al.* [55, 56] and applied to a variety of problems in mechanics and physics. Essentially the same process was advanced simultaneously by Bank and Weiser [34, 37], who focused on the applications to scalar elliptic problems in two dimensions and provided a mathematical analysis of the method. The paper of Bank and Weiser [37] also involved a number of basic ideas that proved to be fundamental to certain theories of *a posteriori* error estimation, including the saturation assumption and the equilibration of boundary data in the context of piecewise linear approximation on triangles.

During the early 1980s the search for effective adaptive methods led to a wide variety of *ad hoc* error estimators. Many of these were based on *a priori* or interpolation estimates, that provided crude but effective indications of features of error sufficient to drive adaptive processes. In this context, we mention the interpolation error estimates of Demkowicz *et al.* [53]. In computational fluid dynamics calculations, these crude interpolation estimates proved to be useful for certain problems in inviscid flow (see Peraire *et al.* [96]), where solutions featured surfaces of discontinuity, shocks, and rarefaction waves. Relatively crude error estimates are sufficient to locate regions in the domain in which discontinuities appear, and these are satisfactory for use as a basis for certain adaptive schemes. However, when more complex features of the solution are present, such as boundary layers or shock-boundary layer interactions, these cruder methods are often disastrously inaccurate.

Zienkiewicz and Zhu [121] developed a simple error estimation technique that is effective for some classes of problems and types of finite element approximations. Their method falls into the category of *recovery-based methods*: Gradients of solutions obtained on a particular partition are smoothed and then compared with the gradients of the original solution to assess error. This approach was modified [122, 123], leading to the *superconvergent patch recovery* method.

Extrapolation methods have been used effectively to obtain global error estimates for both the h and p versions of the finite element method. For example, by using sequences of hierarchical p version approximations, Szabo [106] obtained efficient *a posteriori* estimators for two dimensional linear elasticity problems.

By the early 1990s the basic techniques of *a posteriori* error estimation were established. Attention then shifted to the application to general classes of problems. Verfürth [109] obtained two-sided bounds and derived error estimates for the Stokes problem and the Navier–Stokes equations. An important paper on explicit error residual methods for broad classes of boundary value problems, including nonlinear problems, was presented by Baranger and El-Amri [38]. Estimators for mixed finite element approximation of elliptic problems have been presented in references [43, 48, 49], and Eriksson and

Johnson [64, 65, 66, 72] derived *a posteriori* error estimates for both parabolic and hyperbolic problems.

Most studies have dealt with *a posteriori* error estimation for the h version of the finite element method. The element residual method is applicable to both p and h–p version finite element approximations. An extensive study of error residual methods is reported in the paper by Oden *et al.* [88]. These techniques were applied to nonuniform h–p meshes. Later, in a series of papers, Ainsworth and Oden [11] produced extensions of the element residual method in conjunction with equilibrated boundary data. This was extended to elliptic boundary value problems, elliptic systems, variational inequalities, and indefinite problems such as the Stokes problem and the steady Navier Stokes equations with small data.

The subject of *a posteriori* error estimation for finite element approximation has now reached maturity. The emphasis has now shifted from the development of new techniques to the study of robustness of existing estimators and identifying limits on their performance. Noteworthy in this respect is the work of Babuška *et al.* [29, 31] who conducted an extensive study of the performance and robustness of the main error estimation techniques applied to first-order finite element approximation.

Most of the error estimates developed prior to the mid-1990s pertained to global bounds in energy norms. The late 1990s saw significant advances in extending the theory to local estimates, including estimates of local *quantities of interest* that are crucial in applications. The work on local estimates has occurred with the realization that the error at a particular point of interest in the domain can be *polluted* by errors generated far outside the neighborhood of the point of interest. Such local and pollution errors are discussed in Chapter 8 along with techniques for obtaining both upper and lower bounds on the error in the quantity of interest.

The literature on *a posteriori* error estimation for finite element approximation is vast. We have resisted the temptation to produce an exhaustive survey. The availability of computer databases means that anyone can generate an up-to-date survey with minimal effort. Instead, the bibliography consists primarily of key references and work having a direct influence on our exposition. Surveys of the earlier literature will be found in Ewing [67], Noor and Babuška [86], Oden and Demkowicz [87], and Verfürth [112].

1.3 FINITE ELEMENT NOMENCLATURE

This section is primarily concerned with establishing notation. It is assumed that the reader is familiar with the standard theory of finite element analysis as described in the books of Brenner and Scott [45], Ciarlet [50], and Oden and Reddy [93].

1.3.1 Sobolev Spaces

Let ω be a bounded domain in $\mathbb{R}^2$ with Lipschitz boundary $\partial\omega$. For a nonnegative integer m and any $r \in [1, \infty]$, the Sobolev space $W^{m,r}(\omega)$ is defined in the usual way

$$W^{m,r}(\omega) = \left\{ v \in L^r(\omega) : \frac{\partial^{|\alpha|} v}{\partial \boldsymbol{x}^{\alpha}} \in L^r(\omega), |\boldsymbol{\alpha}| \le m \right\} \tag{1.1}$$

and is equipped with two alternative seminorms

$$|v|_{W^{m,r}(\omega)} = \left[\sum_{|\alpha|=m} \left\| \frac{\partial^{|\alpha|} v}{\partial \boldsymbol{x}^{\alpha}} \right\|_{L^r(\omega)}^r \right]^{1/r} \tag{1.2}$$

and

$$[v]_{W^{m,r}(\omega)} = \left[\left\| \frac{\partial^m v}{\partial x^m} \right\|_{L^r(\omega)}^r + \left\| \frac{\partial^m v}{\partial y^m} \right\|_{L^r(\omega)}^r \right]^{1/r} \tag{1.3}$$

with the norm defined by

$$\|v\|_{W^{m,r}(\omega)} = \left[\sum_{|\alpha|\le m} \left\| \frac{\partial^{|\alpha|} v}{\partial \boldsymbol{x}^{\alpha}} \right\|_{L^r(\omega)}^r \right]^{1/r} . \tag{1.4}$$

The usual modifications are made in the case $r = \infty$. Moreover, the space $W^{m,2}$ will generally be written in the shorthand form H^m. The Sobolev spaces may be extended to noninteger values of m using Hilbert space interpolation [39, 81].

It will be useful to introduce various polynomial type spaces. In particular, for a nonnegative integer p, the space $\mathbb{P}_p$ consists of polynomials of *total* degree at most p in the variables x and y,

$$\mathbb{P}_p = \text{span}\left\{ x^l y^m, 0 \le l, m, l+m \le p \right\} \tag{1.5}$$

while $\mathbb{Q}_p$ consists of polynomials of degree at most p in *each* variable,

$$\mathbb{Q}_p = \text{span}\left\{ x^l y^m, 0 \le l, m \le p \right\} . \tag{1.6}$$

These spaces coincide if $p = 0$ but are otherwise different. A central role will be played by the approximation properties of these spaces.

Theorem 1.1 *Let ω be a bounded, open, connected subset of $\mathbb{R}^2$ with Lipschitz boundary. For each nonnegative integer p and $r \in [1, \infty]$, there exists a constant C such that for all $v \in W^{p+1,r}(\omega)$,*

$$\inf_{w \in \mathbb{P}_p} \|v - w\|_{W^{p+1,r}(\omega)} \le C\, |v|_{W^{p+1,r}(\omega)} \tag{1.7}$$

and

$$\inf_{w \in \mathbb{Q}_p} \|v - w\|_{W^{p+1,r}(\omega)} \leq C\,[v]_{W^{p+1,r}(\omega)}\,. \tag{1.8}$$

Proof. See Theorem 3.1.1 and Exercise 3.1.1 in Ciarlet [50]. ■

Frequently in the later text, we shall assert that a result follows *by a scaling argument.* This is a standard tool in the armory of any finite element analyst and is described in several texts, such as the book of Brenner and Scott [45]. At this point, we may illustrate the basic ingredients of a scaling argument in ascertaining the dependence of the constant C in Theorem 1.1 on the diameter of the domain ω. The diameter d of the domain ω is defined to be

$$d = \sup_{\boldsymbol{x},\boldsymbol{y} \in \omega} \|\boldsymbol{x} - \boldsymbol{y}\|\,. \tag{1.9}$$

The first step is to introduce the scaled domain $\widehat{\omega}$ defined by

$$\widehat{\omega} = \left\{\widehat{\boldsymbol{x}} \in \mathbb{R}^2 : d\widehat{\boldsymbol{x}} \in \omega\right\}. \tag{1.10}$$

Let $v \in W^{m,p}(\omega)$ be given and define the function $\widehat{v} \in W^{m,p}(\widehat{\omega})$ by the rule

$$\widehat{v}(\widehat{\boldsymbol{x}}) = v(\boldsymbol{x}), \quad \widehat{\boldsymbol{x}} \in \widehat{\omega} \tag{1.11}$$

where $\boldsymbol{x} = d\widehat{\boldsymbol{x}}$. Obviously, $\boldsymbol{x} \in \omega$ and so $\widehat{v}$ is well-defined.

Corollary 1.2 *Under the same hypotheses of Theorem 1.1, there exists a constant C that is independent of the diameter d such that*

$$\inf_{w \in \mathbb{P}_p} \sum_{m=0}^{p+1} d^m\,|v - w|_{W^{m,r}(\omega)} \leq C d^{p+1}\,|v|_{W^{p+1,r}(\omega)} \tag{1.12}$$

and

$$\inf_{w \in \mathbb{Q}_p} \sum_{m=0}^{p+1} d^m\,|v - w|_{W^{m,r}(\omega)} \leq C d^{p+1}\,[v]_{W^{p+1,r}(\omega)}\,. \tag{1.13}$$

Proof. Let $v \in W^{m,p}(\omega)$ be given and define $\widehat{v}$ by the rule (1.11). Applying Theorem 1.1 shows that there exists a constant C, depending only on the shape of the domain $\widehat{\omega}$ such that

$$\inf_{\widehat{w} \in \mathbb{Q}_p} \|\widehat{v} - \widehat{w}\|_{W^{p+1,r}(\widehat{\omega})} \leq C\,[\widehat{v}]_{W^{p+1,r}(\widehat{\omega})}\,. \tag{1.14}$$

The important point to notice here is that the constant C is independent of the diameter of the domain ω. Let $\widehat{w} \in \mathbb{Q}_p$ be a polynomial to be chosen later and define $w \in \mathbb{Q}_p$ by the rule

$$w(\boldsymbol{x}) = \widehat{w}(\widehat{\boldsymbol{x}}), \quad \widehat{\boldsymbol{x}} \in \widehat{\omega}.$$

Now, applying the change of variable $\widehat{\boldsymbol{x}} = d^{-1}\boldsymbol{x}$ easily leads to the conclusion

$$|v - w|^r_{W^{m,r}(\omega)} = d^{2-mr}\,|\widehat{v} - \widehat{w}|^r_{W^{m,r}(\omega)}$$

for $m = 0, \ldots, p+1$. In the same fashion, it is readily shown that

$$[\widehat{v}]^r_{W^{p+1,r}(\widehat{\omega})} = d^{-2+r(p+1)}\,[v]^r_{W^{p+1,r}(\omega)}\,.$$

These results imply that

$$|v - w|^r_{W^{m,r}(\omega)} \leq d^{2-mr}\,|\widehat{v} - \widehat{w}|^r_{W^{m,r}(\widehat{\Omega})} \leq d^{2-mr}\,\|\widehat{v} - \widehat{w}\|^r_{W^{p+1,r}(\Omega)}$$

and then, taking the infimum over $\widehat{w} \in \mathbb{Q}_p$ and applying (1.14) shows that

$$\inf_{w\in\mathbb{Q}_p} |v - w|^r_{W^{m,r}(\omega)} \leq Cd^{2-mr}\,[\widehat{v}]^r_{W^{p+1,r}(\widehat{\omega})} \leq Cd^{2-mr-2+r(p+1)}\,[v]^r_{W^{p+1,r}(\omega)}\,.$$

In each instance, the constant C may take different values, but the values are independent of the diameter d of the domain ω. In summary, we have shown that for all $m = 0, \ldots, p+1$, there holds

$$\inf_{w\in\mathbb{Q}_p} \sum_{m=0}^{p+1} d^m\,|v - w|_{W^{m,r}(\omega)} \leq Cd^{p+1}\,[v]_{W^{p+1,r}(\omega)}$$

where the constant C depends on the shape of the domain ω and on the integers m and p, but not on the diameter d. The remaining assertion follows using analogous arguments. ∎

It is also worthwhile noting that the role of the generic constant C in this particular proof is typical of many arguments in the analysis of finite element methods in general and of several of the later arguments in this text. As was pointed out, the particular value of C need not be the same in any two places, even in the same equation. Strictly speaking, one should perhaps label the various constants $C_1, C_2, \ldots$. However, at the end of the day, what remains is nonetheless a generic constant whose value is not explicitly known. Thus, in order to avoid a proliferation of subscripts, we shall follow the common practice of not distinguishing between different unknown constants C_{42} and C_{1066}, say, and instead prefer to denote both by the letter C.

1.3.2 Inverse Estimates

Scaling arguments such as the one used to prove Corollary 1.2 may also be used to obtain *inverse estimates.* It follows from the definition of the Sobolev spaces that if a function v belongs to the space $W^{m,r}(\omega)$, then it automatically belongs to the space $W^{k,r}(\omega)$, $k \leq m$, and there exists a constant C such that

$$\|v\|_{W^{k,r}(\omega)} \leq C\,\|v\|_{W^{m,r}(\omega)}\,.$$

The reverse inequality is not true in general. However, if the function v belongs to a finite-dimensional subspace of $W^{m,r}(\omega)$, then such an *inverse inequality* does hold, with a constant C depending on the diameter of the domain ω. The precise result reads as follows:

Theorem 1.3 *Let $\omega \subset \mathbb{R}^n$ be a bounded domain, with diameter d. Let $r \in [1, \infty]$ and $m \geq 0$. Suppose that $P \subset W^{m,r}(\omega)$ is a finite-dimensional subspace. Then, for all $v \in P$ and $0 \leq t \leq s \leq m$, there holds*

$$|v|_{W^{s,r}(\omega)} \leq Cd^{-(s-t)} |v|_{W^{t,r}(\omega)} \tag{1.15}$$

where C is a positive constant depending on m, r, s, t, the space P, and the subdomain ω.

Proof. Let $\widehat{\omega}$ be the scaled domain defined in (1.10) and, for $v \in W^{k,r}(\omega)$, define the function $\widehat{v}$ by the rule (1.11). The finite-dimensional space $\widehat{P}$ is defined to be $\widehat{P} = \{\widehat{v} : v \in P\}$. Let $v \in P$, then a simple change of variable shows that

$$|\widehat{v}|_{W^{k,r}(\widehat{\omega})} = d^{k-n/r} |v|_{W^{k,r}(\omega)}$$

for $0 \leq k \leq m$. By the equivalence of norms on the finite-dimensional space $\widehat{P}$, there exists a constant C, independent of d, such that for all $\widehat{v} \in \widehat{P}$,

$$\|\widehat{v}\|_{W^{k,r}(\widehat{\omega})} \leq C \|\widehat{v}\|_{L_r(\widehat{\omega})} .$$

Hence, since

$$d^{k-n/r} |v|_{W^{k,r}(\omega)} = |\widehat{v}|_{W^{k,r}(\widehat{\omega})} \leq \|\widehat{v}\|_{W^{k,r}(\widehat{\omega})}$$

and

$$\|\widehat{v}\|_{L_r(\widehat{\omega})} = d^{-n/r} \|v\|_{L_r(\omega)} ,$$

it follows that

$$|v|_{W^{k,r}(\omega)} \leq Cd^{-k} \|v\|_{L_r(\omega)} \quad \forall v \in P.$$

This proves the result in the case $t = 0$. The same method of proof may be used to show that this estimate holds, possibly with a different constant C, for all functions v belonging to the finite-dimensional space

$$P^{(t)} = \{D^{\alpha} v : v \in P, |\alpha| = t\}$$

for each $0 \leq t \leq m - k$. In other words, it follows that for each $w \in P$,

$$|D^{\alpha} w|_{W^{k,r}(\omega)} \leq Cd^{-k} \|D^{\alpha} w\|_{L_r(\omega)} \leq Cd^{-k} |w|_{W^{t,r}(\omega)}$$

for $|\alpha| = t$. Since α is otherwise arbitrary, it follows that

$$|w|_{W^{k+t,r}(\omega)} \leq Cd^{-k} |w|_{W^{t,r}(\omega)} ;$$

hence, choosing $k = s - t$ gives the result claimed. ■

1.3.3 Finite Element Partitions

The polynomial spaces $\mathbb{P}_p$ and $\mathbb{Q}_p$ are used to construct finite elements on triangular and quadrilateral elements, respectively. The next section is concerned with describing the precise assumptions on the meshes and the construction of the finite element spaces.

Let Ω be a polygonal domain with boundary $\partial\Omega$. A finite element partition $\mathcal{P}$ of Ω is a collection $\{K\}$ of elements such that:

1. The elements form a partition of the domain, that is, $\overline{\Omega} = \bigcup_{K \in \mathcal{P}} \overline{K}$.

2. Each element is a triangle or convex quadrilateral contained in Ω.

3. The nonempty intersection of (the closure) of each distinct pair of elements is either a single common vertex or a single common edge of both elements.

The diameter of a triangle K is denoted by h_K and the diameter of the largest circle that may be inscribed in K is denoted by ρ_K. The regularity of the triangle is measured by the ratio

$$\kappa_K = \frac{h_K}{\rho_K}. \tag{1.16}$$

A similar concept may be defined for a quadrilateral element through a modified procedure. Let $\boldsymbol{a}_l$, $l = 1, \ldots, 4$, denote the vertices of the quadrilateral enumerated anti-clockwise. Let T_l denote the triangle formed from vertices $\boldsymbol{a}_l$, $\boldsymbol{a}_{l+1}$, $\boldsymbol{a}_{l+2}$—the indices being counted modulo four. Let h_l denote the diameter of T_l and let ρ_l denote the diameter of the largest inscribed ball and define

$$h_K = \max_l h_l \text{ and } \rho_K = \min_l \rho_l, \tag{1.17}$$

then the regularity of the quadrilateral is measured by the ratio (1.16).

A partitioning of Ω is said to be *regular* if there exists a constant κ such that $\kappa_K \leq \kappa$ for all elements in the partition. More generally, regular families of partitions consist of partitions comprised of elements for which the parameter κ_K is uniformly bounded over the whole family.

This property is sometimes referred to in the literature as the shape regularity of the elements. It is important to realize that the assumption of regularity permits partitions of the domain Ω into meshes that may be very highly locally refined containing elements of quite different sizes. In particular, the regularity assumption does not rule out the types of meshes that arise in an adaptive refinement procedure: Elements where the solution is nonsmooth may be refined at every step of the procedure, while other elements may very well not be refined at all during the process.

1.3.4 Finite Element Spaces on Triangles

Let p be a nonnegative integer and let $\mathcal{P}$ be a regular partition of the domain Ω into triangular elements. The finite element subspace of order p associated with the partition $\mathcal{P}$ is defined by

$$X = \left\{ v \in C(\overline{\Omega}) : \forall K \in \mathcal{P}, v|_K \in \mathbb{P}_p \right\}. \tag{1.18}$$

The degrees of freedom of the space X may be identified with function evaluations at *nodes*. The set of nodes, $\{\boldsymbol{x}_k : k \in \mathcal{N}\}$, is assumed to consist of the vertices of the principal lattice on each of the elements and, as such, includes the vertices of the elements. A Lagrange basis for the space X consists of functions $\{\theta_k : k \in \mathcal{N}\}$ satisfying the following conditions:

1. $\theta_k \in X$ for all $k \in \mathcal{N}$;

2. $\theta_k(\boldsymbol{x}_l) = \delta_{kl}$ for all $k, l \in \mathcal{N}$ where δ_{kl} is the Kronecker symbol.

Often, such a basis is referred to as a *nodal basis* in the engineering finite element literature. While such a basis is convenient for the purposes of exposition, it is generally recognized that alternative bases are more appropriate for the purposes of practical implementation, especially in the case of high-order elements.

In the implementation and analysis of finite element methods, an important role is played by *reference elements*. The reference element for triangular elements may be chosen to be

$$\widehat{K} = \{(\widehat{x}, \widehat{y}) : 0 \leq \widehat{x} \leq 1, 0 \leq \widehat{y} \leq 1 - \widehat{x}\}. \tag{1.19}$$

Each triangular element K may be regarded as the image of the reference triangle $\widehat{K}$ under a continuous, affine, invertible transformation $\mathcal{F}_K$. The transformation may be written in the form

$$\mathcal{F}_K(\widehat{\boldsymbol{x}}) = \boldsymbol{A}_K \widehat{\boldsymbol{x}} + \boldsymbol{b}_K \tag{1.20}$$

where $\boldsymbol{A}_K$ is a matrix and $\boldsymbol{b}_K$ is a vector. The regularity of the element K means that the $\boldsymbol{A}_K$ is nonsingular, and there exists a constant C that depends only on the regularity parameter κ_K such that

$$\left. \begin{aligned} \|\boldsymbol{A}_K\| &\leq C h_K \\ \|\boldsymbol{A}_K^{-1}\| &\leq C \rho_K^{-1} \\ C\rho_K^2 \leq |\det(\boldsymbol{A}_K)| &\leq C h_K^2 \end{aligned} \right\} \tag{1.21}$$

where $\|\cdot\|$ is a matrix norm.

1.3.5 Finite Element Spaces on Quadrilaterals

The construction of the finite element subspace of order p on a regular partitioning of Ω into quadrilateral elements is based on the polynomial space $\mathbb{Q}_p$. The main difference compared with finite element spaces on triangles is that the actual functions are no longer necessarily polynomials. Instead, so-called *pull-back* functions are used. In order to describe these functions more precisely, we first introduce the reference domain corresponding to quadrilateral elements to be the square,

$$\widehat{K} = \{(\widehat{x}, \widehat{y}) : -1 \leq \widehat{x}, \widehat{y} \leq 1\}. \tag{1.22}$$

As for triangular elements, each quadrilateral element K may be regarded as being the image of the reference domain under a transformation $\mathcal{F}_K : \widehat{K} \to K$, where $\mathcal{F}_K$ is an invertible, bilinear mapping such that the vertices of the element K are the images of the vertices of the reference element with their orientation preserved. The mapping $\mathcal{F}_K$ possesses properties similar to the case of triangular elements. However, unless the element happens to be a parallelogram, the mapping will not be affine, and the entries of the corresponding matrix $\boldsymbol{A}_K$ will be functions of the local coordinates $(\widehat{x}, \widehat{y})$. Therefore, the conditions corresponding to (1.21) are formulated slightly differently:

$$\left.\begin{aligned} \|D\mathcal{F}_K\|_{L_\infty(\widehat{K})} &\leq Ch_K \\ \left\|D\mathcal{F}_K^{-1}\right\|_{L_\infty(\widehat{K})} &\leq C\kappa_K/\rho_K \\ C\rho_K^2 \leq \|J_K\|_{L_\infty(\widehat{K})} &\leq Ch_K^2 \end{aligned}\right\} \tag{1.23}$$

where J_K and $D\mathcal{F}_K$ denote, respectively, the Jacobian and the Jacobian matrix of the transformation $\mathcal{F}_K$, and the norm is now the supremum of the matrix norms over the reference element.

The finite element subspace on a regular quadrilateral mesh is defined to be

$$X = \left\{v \in C(\overline{\Omega}) : \forall K \in \mathcal{P}, v = \widehat{v} \circ \mathcal{F}_K^{-1}; \widehat{v} \in \mathbb{Q}_p\right\}. \tag{1.24}$$

Analogously to the case of triangles, the degrees of freedom of the space X are associated with function evaluations at the nodes $\{\boldsymbol{x}_k : k \in \mathcal{N}\}$, where the nodes on each element K are associated with the images of the vertices of a regularly spaced lattice on the reference element, under the mapping $\mathcal{F}_K$. As before, the set of nodes includes the vertices of the elements. A Lagrange basis $\{\theta_k : k \in \mathcal{N}\}$ for the space X is defined satisfying precisely the same conditions as in the case of triangular elements. However, the basis functions for quadrilateral meshes will generally be pull-back functions as opposed to polynomials for triangular meshes. These ideas may be readily extended to hybrid meshes consisting of both quadrilateral and triangular elements.

1.3.6 Properties of Lagrange Basis Functions

This section is concerned with presenting some elementary properties of the Lagrange basis that will be useful in the sequel:

Lemma 1.4 *Let X be a finite element subspace of order p constructed on a regular partition of Ω, and let $\{\theta_k : k \in \mathcal{N}\}$ be a Lagrange basis. Then, there exists a constant C depending on the polynomial degree p and the regularity of the partition such that*

$$\|\theta_k\|_{L_\infty(K)} \leq C \tag{1.25}$$

and

$$\|\nabla\theta_k\|_{L_\infty(K)} \leq Ch_K^{-1} \tag{1.26}$$

where $K \in \mathcal{P}$ is any element.

Proof. Let K be any element over which θ_k is not identically zero. Then, K is the image of a reference element $\widehat{K}$ under the transformation $\mathcal{F}_K$. Hence, by virtue of the construction of the space X, it follows that for any $\boldsymbol{x} \in K$,

$$\theta_k(\boldsymbol{x}) = \widehat{\theta}(\mathcal{F}_K^{-1}(\boldsymbol{x}))$$

where $\widehat{\theta}$ belongs to the space $\mathbb{P}_p$ or $\mathbb{Q}_p$ depending on whether K is a triangle or a quadrilateral. Consequently,

$$|\theta_k(\boldsymbol{x})| \leq \left\|\widehat{\theta}\right\|_{L_\infty(\widehat{K})} \leq C$$

where C depends on the geometry of the reference element and the polynomial degree p. Furthermore,

$$|\nabla\theta_k(\boldsymbol{x})| \leq \left\|\widehat{\nabla}\widehat{\theta}\right\|_{L_\infty(\widehat{K})} \|D\mathcal{F}_K^{-1}\|_{L_\infty(\widehat{K})}$$

where $\widehat{\nabla}$ denotes the gradient with respect to the coordinates on the reference element. The result then follows thanks to (1.16) along with either (1.21) or (1.23) as appropriate. ∎

1.3.7 Finite Element Interpolation

Let $\mathcal{P}$ be a regular partitioning of the domain Ω into triangular or quadrilateral elements, and for a nonnegative integer p let X denote the finite element subspace constructed on the partition. The X-interpolant, $I_X u$, of a continuous function $u \in C(\overline{\Omega})$ is uniquely defined by the conditions

$$I_X u \in X : \quad I_X u(\boldsymbol{x}_k) = u(\boldsymbol{x}_k), \forall k \in \mathcal{N} \tag{1.27}$$

where $\{\boldsymbol{x}_k : k \in \mathcal{N}\}$ are the nodes of the partition. An explicit representation for the interpolant may be given in terms of the Lagrange basis functions

$$I_X u = \sum_{k \in \mathcal{N}} u(\boldsymbol{x}_k)\theta_k. \tag{1.28}$$

The interpolant is a useful theoretical tool since it provides a quasi-optimal approximation from the finite element subspace. It is therefore of interest to develop approximation theoretic results for the interpolant. The following classic result may be found in any standard finite element text.

Theorem 1.5 *Let $r \in [1, \infty]$ and, for a nonnegative integer p, let X denote the finite element subspace constructed on a regular partition $\mathcal{P}$ of Ω into triangular or quadrilateral elements. Let $s \in [0,1]$ and t satisfy $s \le t \le p+1$. Then, there exists a constant C (depending only on the regularity κ of the partition) such that for all $u \in W^{t,r}(\Omega)$, the X-interpolant satisfies*

$$|u - I_X u|_{W^{s,r}(K)} \le C h_K^{t-s} |u|_{W^{t,r}(K)} \tag{1.29}$$

for all elements $K \in \mathcal{P}$, provided that $p + 1 > 2/r$ if $r > 1$ or provided that $p \ge 1$ if $r = 1$.

Proof. See Ciarlet [50], Brenner and Scott [45] or Oden and Reddy [93]. ∎

1.3.8 Patches of Elements

For the purposes of *a posteriori* error analysis, it will be worthwhile to consider properties of certain local patches of elements. Let $\mathcal{P}$ be a regular partition of Ω into triangles or quadrilaterals and let K be any element in the partition. The subdomain $\widetilde{K}$ consists of the element K along with the elements sharing at least one common vertex with K:

$$\widetilde{K} = \text{int}\left\{ \bigcup K', K' \in \mathcal{P} : \overline{K}' \cap \overline{K} \text{ is nonempty} \right\}. \tag{1.30}$$

The subdomain $\widetilde{K}$ is referred to as the *patch* associated with element K. The reader may find it helpful at this stage to note that the notation $\widetilde{\bullet}$ will be employed throughout to denote a patch of elements associated with an entity $\bullet$.

The dimensions of the patch $\widetilde{K}$ are measured in terms of the quantities

$$h_{\widetilde{K}} = \max_{K' \subseteq \widetilde{K}} h_{K'} \tag{1.31}$$

and

$$\rho_{\widetilde{K}} = \min_{K' \subseteq \widetilde{K}} \rho_{K'}, \tag{1.32}$$

while the regularity of the patch is measured by

$$\kappa_{\widetilde{K}} = \frac{h_{\widetilde{K}}}{\rho_{\widetilde{K}}}. \tag{1.33}$$

The regularity of the partition $\mathcal{P}$ is inherited by the patches.

Theorem 1.6 *Let $\mathcal{P}$ be a regular partition of Ω into triangles or quadrilaterals, and let $\widetilde{K}$ denote the patch associated with a particular element $K \in \mathcal{P}$. Then, there exists a constant C, depending only on κ, such that*

$$h_{\widetilde{K}} \leq C h_{K'} \quad \forall K' \subset \widetilde{K} \tag{1.34}$$

and

$$\kappa_{\widetilde{K}} \leq C \kappa_{K'} \quad \forall K' \subset \widetilde{K}. \tag{1.35}$$

Moreover, the number of elements in each patch and the number of patches containing a particular element are bounded by a constant depending only on κ.

Proof. The results follow easily from geometrical considerations. See, for example, reference [51]. ∎

The conditions (1.34)–(1.35) are referred to as *quasi-uniformity* conditions in the finite element literature. More generally, if the set of elements belonging to a family of partitions of Ω satisfy conditions (1.34)–(1.35), then the family is said to be *quasi-uniform.* The assumption of quasi-uniformity for a family of partitions is often made in the classical texts on finite elements. However, it should be borne in mind that the assumption rules out families of partitions that typically arise in an adaptive refinement procedure. For this reason, the assumption of quasi-uniformity is avoided in the present work. However, the above results show that the regularity assumption on the elements in the partition implies that the elements contained in a particular patch are *quasi-uniform* even though the entire set of elements might not be quasi-uniform.

1.3.9 Regularized Approximation Operators

In the analysis of *a posteriori* error estimators for second-order elliptic boundary value problems, one is generally dealing with approximations to solutions of problems that may not possess a significant additional smoothness beyond the $H^1(\Omega)$ regularity guaranteed by the Lax–Milgram Theorem. Frequently, one is faced with the desire to construct an accurate approximation to the true solution $u \in H^1(\Omega)$ from the finite element subspace. Therefore, if the analysis is to apply to situations where the true solution u may in principle only belong to the space $H^1(\Omega)$, then it is necessary to employ more sophisticated approximation operators than the classical interpolation operators that require additional smoothness properties of the function u to be interpolated in order to guarantee that the pointwise values of the function u are well-defined.

The following result presents a regularized finite element approximation operator that does not require additional smoothness of the function u.

Theorem 1.7 *Let $r \in [1, \infty)$, and for a nonnegative integer p let X denote the finite element subspace constructed on a regular partition $\mathcal{P}$ of the polygonal domain Ω into triangular or quadrilateral elements. Let $s \in [0, 1]$*

and t satisfy $s \leq t \leq p+1$. Then, there exists a bounded, linear operator $\mathcal{I}_X : W^{t,r}(\Omega) \longrightarrow X$ and a constant C (depending only on the regularity κ of the partition) such that for all $u \in W^{t,r}(\Omega)$ and all elements $K \in \mathcal{P}$ we have

$$|u - \mathcal{I}_X u|_{W^{s,r}(K)} \leq C h_K^{t-s} |u|_{W^{t,r}(\widetilde{K})} \tag{1.36}$$

and

$$|u - \mathcal{I}_X u|_{W^{s,r}(\gamma)} \leq C h_K^{t-s-1/r} \|u\|_{W^{t,r}(\widetilde{K})} \tag{1.37}$$

where γ is any edge of the element K and $\widetilde{K}$ denotes the patch of elements associated with K.

Proof. See Theorem 4.4 and Remark 8 in Bernardi and Girault [40]. ■

1.4 MODEL PROBLEM

Let $\Omega \subset \mathbb{R}^2$ be a bounded domain with a Lipschitz boundary $\partial\Omega$. Consider the model elliptic boundary value problem of finding the solution u of

$$-\Delta u + cu = f \text{ in } \Omega \tag{1.38}$$

subject to the boundary conditions

$$\frac{\partial u}{\partial n} = g \text{ on } \Gamma_N \tag{1.39}$$

and

$$u = 0 \text{ on } \Gamma_D. \tag{1.40}$$

The data are assumed to be smooth, that is, $f \in L_2(\Omega)$, $g \in L_2(\Gamma_N)$, c is a nonnegative constant, and the boundary segments Γ_N and Γ_D are assumed to be disjoint with $\overline{\Gamma}_N \cup \overline{\Gamma}_D = \partial\Omega$. The unit outward normal vector to $\partial\Omega$ is denoted by $\boldsymbol{n}$ and belongs to the space $[L_\infty(\partial\Omega)]^n$.

The variational form of this problem is to find $u \in V$ such that

$$B(u, v) = L(v) \quad \forall v \in V \tag{1.41}$$

where V is the space

$$V = \{v \in H^1(\Omega) : v = 0 \text{ on } \Gamma_D\} \tag{1.42}$$

and where

$$B(u, v) = \int_\Omega (\boldsymbol{\nabla} u \cdot \boldsymbol{\nabla} v + cuv) \, d\boldsymbol{x} \tag{1.43}$$

and

$$L(v) = \int_\Omega f v \, d\boldsymbol{x} + \int_{\Gamma_N} g v \, ds. \tag{1.44}$$

Here, and in the remainder of the text, conditions of the type $v = 0$ on Γ_D are to be understood in the sense of traces [81].

Suppose that $X \subset V$ is a finite element subspace. The finite element approximation of this problem is to find $u_X \in X$ such that

$$B(u_X, v_X) = L(v_X) \quad \forall v_X \in X. \tag{1.45}$$

The error $e = u - u_X$ belongs to the space V and satisfies

$$B(e, v) = B(u, v) - B(u_X, v) = L(v) - B(u_X, v) \quad \forall v \in V. \tag{1.46}$$

Moreover, the standard orthogonality condition for the error in the Galerkin projection holds

$$B(e, v_X) = 0 \quad \forall v_X \in X. \tag{1.47}$$

1.5 PROPERTIES OF A POSTERIORI ERROR ESTIMATORS

There are many techniques for error estimation. One can extrapolate approximate solutions obtained on sequences of progressively finer meshes or on sequences of meshes with shape functions of increasing order and then compare solutions to obtain an indication of the error. Such methods can be quite effective when data structures admit such multilevel computations. One popular method amongst the engineering community is to post-process the approximation u_X to obtain more accurate representations of the gradient $G(u_X)$. One can then use the difference $G(u_X) - \nabla u_X$ as an estimate for the error. This type of approach can lead to surprisingly good error estimators and is discussed in Chapter 4. One of the weaknesses of the method and at the same time, one of its advantages, is that no use is made of the information from the original problem.

Other error estimators make use of the data of the problem and properties of the error in various ways. For instance, the approximation error satisfies the residual equation (1.46) and the orthogonality condition (1.47). A residual equation similar to (1.46) may be obtained by integrating by parts over each element leading to

$$\int_K (\nabla e \cdot \nabla v + cev)\,\mathrm{d}\boldsymbol{x} = \int_K rv\,\mathrm{d}\boldsymbol{x} + \oint_{\partial K} v\,\boldsymbol{n}_K \cdot \nabla e|_K\,\mathrm{d}s \tag{1.48}$$

where r is the *residual*

$$r = f + \Delta u_X - cu_X \tag{1.49}$$

and $\boldsymbol{n}_K$ is the unit exterior normal to the element boundary ∂K. Under suitable conditions, the solution to (1.48) may be bounded by

$$|\!|\!| e |\!|\!|_K \leq C_1 \, \|r\|_{L_2(K)} + C_2 \, \|R\|_{L_2(\partial K)} \tag{1.50}$$

where R is an approximation to $\boldsymbol{n}_K \cdot \boldsymbol{\nabla} e$ on the element boundary, and C_1 and C_2 may depend upon the element size h_K and other mesh parameters. Replacing the boundary fluxes of the true error over the element boundary by a suitable approximation leads to a bound on the error on the element K (apart from the constants C_1 and C_2). This type of estimator is referred to as an *explicit estimator* and is discussed in Chapter 2.

The presence of the constants C_1 and C_2 in the explicit *a posteriori* error estimators leads one to consider trying to solve an approximate local boundary value problem for the error of the form

$$\int_K (\boldsymbol{\nabla}\phi_K \cdot \boldsymbol{\nabla} v + c\phi_K v)\,\mathrm{d}\boldsymbol{x} = \int_K rv\,\mathrm{d}\boldsymbol{x} + \oint_{\partial K} v\,(g_K - \boldsymbol{n}_K \cdot \boldsymbol{\nabla} u_X)\,\mathrm{d}s \quad (1.51)$$

where g_K is an approximation to the boundary flux. The solution ϕ_K may be used as

$$|||\phi_K|||_K^2 = \int_K (|\boldsymbol{\nabla}\phi_K|^2 + c\phi_K^2)\,\mathrm{d}\boldsymbol{x}. \quad (1.52)$$

to provide a measure of the error content in the approximation associated with element K. The approach raises a number of issues:

- The infinite-dimensional space containing the error must be approximated by an appropriate finite-dimensional subspace.
- The boundary flux $\boldsymbol{n}_K \cdot \boldsymbol{\nabla} u$ must be approximated in some effective way.
- If $c = 0$ the error residual problem (1.51) may have no solution unless the compatibility condition

  $$\int_K r\,\mathrm{d}\boldsymbol{x} + \oint_{\partial K} (g_K - \boldsymbol{n}_K \cdot \boldsymbol{\nabla} u_X)\,\mathrm{d}s = 0 \quad (1.53)$$

  is satisfied.

The general process just described is an example of an *implicit error residual method.* It is said to be implicit because the error residual problem must be solved over each element to determine the *error estimator* $|||\phi_K|||_K$. Implicit estimators are the subject of Chapter 3. Chapter 6 deals with implicit error estimators in which the boundary fluxes are specially constructed so that the local problem is well-posed. The resulting estimators will be shown to possess a number of important properties.

If η_K is a local error estimator on element K then the global error estimate η is usually taken to be

$$\eta = \left\{ \sum_{K \in \mathcal{P}} \eta_K^2 \right\}^{1/2}. \quad (1.54)$$

A major property demanded of all successful error estimators is that positive constants C_1 and C_2 exist such that

$$C_1 \|e\| \leq \eta \leq C_2 \|e\| \tag{1.55}$$

where $\|e\|$ is the global error in the energy norm. Then η tends to zero at the same *rate* as the true error. The quality of an estimator is often judged by global *effectivity indices*

$$\theta = \frac{\eta}{\|e\|} \tag{1.56}$$

or *local effectivity indices*

$$\theta_K = \frac{\eta_K}{\|e\|_K}. \tag{1.57}$$

These indices can be used to measure the quality of an estimator when the exact error or a good approximation of it are known. Naturally, one hopes that effectivity indices close to unity can be obtained, but global effectivity indices of 2.0–3.0 or even higher are often regarded as acceptable in many engineering applications.

Throughout, the ideas are presented for the simple model problem in the plane. For the most part, the analysis may be easily extended to three dimensions. Therefore, we shall comment on higher dimensions only in cases where the extension is not immediately apparent. The extension of the results to more general problems may be less straightforward. Therefore, in Chapter 9, applications to more complicated problems are given, including problems with side constraints (such as the Stokes' problem), unsymmetric operators (Oseen's equations), and nonlinearities (the incompressible Navier–Stokes' equations).

It is hoped that by presenting the results in a single notation and framework, the interrelations between different techniques will be more apparent. Chapters 2, 3, and 4 may be read independently. Chapter 6 is also largely independent of the earlier chapters, although the reader might find it helpful to first read Chapter 3.

1.6 BIBLIOGRAPHICAL REMARKS

A regularized approximation operator applicable to triangular elements was developed by Clément [51]. Dupont and Scott [60] obtained approximation results for piecewise polynomials based on regularized Taylor series expansions. Scott and Zhang [102] constructed a regularized approximation operator over triangular elements by incorporating averages of values on element edges, assuming that the function to be approximated was sufficiently smooth for the relevant traces to exist. The interpolation operator for triangular and quadrilateral elements given in the text is taken from Bernardi and Girault [40]. Bounds on the constants appearing in the bounds for such interpolation operators are given in reference [113].

2

Explicit A Posteriori Estimators

2.1 INTRODUCTION

Consider the model problem in Section 1.4 and suppose that the finite element approximation u_X is in hand. The central issue in *a posteriori* error estimation is embodied in the following problem: *Given the Galerkin approximation u_X, the load data f, and the boundary data g, find a quantitative estimate for the true error $e = u - u_X$ measured in a specified norm.*

In seeking an accurate, quantitative estimate for the error, use may be made of the fact that the true error is characterized as the unique solution $e \in V$ of the variational boundary value problem,

$$B(e, v) = B(u, v) - B(u_X, v) = L(v) - B(u_X, v) \quad \forall v \in V. \tag{2.1}$$

Furthermore, since u_X is a Galerkin approximation, the error satisfies the *Galerkin orthogonality property*

$$B(e, v_X) = 0 \quad \forall v_X \in X. \tag{2.2}$$

It will transpire that these two properties together enable one to derive quite accurate estimates for the error, particularly measured in the energy norm, $|||e|||$.

In principle, the error function e itself could be reconstructed as accurately as desired simply by approximating the solution of the boundary value problem (2.1). However, it is clear that this procedure would be equivalent to solving the original problem, so the cost associated with computing a further, more accurate approximation for the purpose of error estimation would either be prohibitively expensive, or the effort involved would be better directed to obtaining an improved Galerkin approximation u_X whose error would presumably then have to be estimated in any case. Thus, in seeking an *a posteriori* error estimation it is often tacitly understood that one should rule out any procedure that entails a computational effort that is of the same order as simply recomputing a more accurate approximation of the original problem.

We begin our study of *a posteriori* error estimators with a description of *explicit* estimators. The term *explicit* is used in a context similar to the characterization of explicit and implicit difference schemes for time-marching algorithms: Explicit schemes involve a direct computation using available data while implicit schemes involve the solution of an algebraic system of equations. An explicit *a posteriori* error estimator employs the residuals in the current approximation directly while an implicit *a posteriori* error estimator uses the residuals indirectly and generally involves the solution of a small linear algebraic system. Implicit estimators will be the subject of later chapters. As might be expected, explicit schemes generally require less computational effort than implicit schemes but involve compromises in robustness and in utility as a means for accurate, quantitative error estimation.

The following section illustrates how the two basic properties of the error expressed in equations (2.1) and (2.2) may be exploited to derive a simple *a posteriori* error estimate for the true error measured in the energy norm. Later, it is shown how estimates may be obtained for the error measured in alternative norms, such as the L_2 (root mean square) or L_∞ (pointwise) norms.

2.2 A SIMPLE A POSTERIORI ERROR ESTIMATE

The first step is to decompose the residual equation (2.1) for the true error into local contributions from each element. The localization of the residual equation to the element level is a key step in circumventing the prohibitive costs entailed in dealing with a global problem, and this basic theme will arise in various guises throughout the derivation of all the *a posteriori* error estimates.

Let $v \in V$ be chosen arbitrarily. Then, writing the single integral over the whole domain as a sum of integrals over the individual elements gives

$$B(e,v) = \sum_{K\in\mathcal{P}} \left\{ \int_K fv \, \mathrm{d}\boldsymbol{x} + \int_{\partial K\cap\Gamma_N} gv \, \mathrm{d}\boldsymbol{x} - \int_K (\boldsymbol{\nabla} u_X \cdot \boldsymbol{\nabla} v + c u_X v) \, \mathrm{d}\boldsymbol{x} \right\}.$$

Applying integration by parts to each of the terms in the final summation and rearranging terms leads to

$$B(e,v) = \sum_{K\in\mathcal{P}} \left\{ \int_K rv \, \mathrm{d}\boldsymbol{x} + \int_{\partial K\cap\Gamma_N} Rv \, \mathrm{d}s - \int_{\partial K\backslash\Gamma_N} \frac{\partial u_X}{\partial n_K} v \, \mathrm{d}s \right\} \tag{2.3}$$

where r is the *interior residual*

$$r = f + \Delta u_X - c u_X \text{ in } K \tag{2.4}$$

and R is the *boundary residual*

$$R = g - \frac{\partial u_X}{\partial n_K} \text{ on } \partial K \cap \Gamma_N \tag{2.5}$$

with $\boldsymbol{n}_K$ being the unit outward normal vector to ∂K. Each of these quantities is well-defined thanks to the smoothness of the data and the regularity of the approximation u_X when restricted to a single element. The contribution from the final term in (2.3) can be rewritten by observing that the (trace of the) function v matches along an edge shared by two elements, giving

$$B(e, v) = \sum_{K \in \mathcal{P}} \left\{ \int_K r v \, \mathrm{d}\boldsymbol{x} + \int_{\partial K \cap \Gamma_N} R v \, \mathrm{d}s \right\} - \sum_{\gamma \in \partial \mathcal{P} \backslash \partial \Omega} \int_\gamma \left[\frac{\partial u_X}{\partial n} \right] v \, \mathrm{d}s, \tag{2.6}$$

where the final summation is over the set $\partial \mathcal{P} \backslash \partial \Omega$ consisting of the interelement edges γ on the interior of the mesh. The quantity

$$\left[\frac{\partial u_X}{\partial n} \right] = \boldsymbol{n}_K \cdot (\boldsymbol{\nabla} u_X)_K + \boldsymbol{n}_{K'} \cdot (\boldsymbol{\nabla} u_X)_{K'} \tag{2.7}$$

defined on the edge γ separating elements K and K' represents the *jump discontinuity* in the approximation to the normal flux on the interface. The identity (2.6) can be written more compactly by extending the definition of the boundary residual to incorporate the jump discontinuity in the flux. Therefore, on interior edges the definition (2.5) is augmented by

$$R = - \left[\frac{\partial u_X}{\partial n} \right] \tag{2.8}$$

so that (2.6) then becomes

$$B(e, v) = \sum_{K \in \mathcal{P}} \int_K r v \, \mathrm{d}\boldsymbol{x} + \sum_{\gamma \in \partial \mathcal{P}} \int_\gamma R v \, \mathrm{d}s \quad \forall v \in V \tag{2.9}$$

where the final summation now extends over all the edges in the partition $\mathcal{P}$.

The orthogonality property (2.2) is brought into play as follows. For given $v \in V$, let $\mathcal{I}_X v$ be the approximation to v from the subspace X defined in Theorem 1.7. Thanks to (2.2) and the identity (2.9), there holds

$$0 = \sum_{K \in \mathcal{P}} \int_K r \mathcal{I}_X v \, \mathrm{d}\boldsymbol{x} + \sum_{\gamma \in \partial \mathcal{P}} \int_\gamma R \mathcal{I}_X v \, \mathrm{d}s \tag{2.10}$$

and then combining this with identity (2.9) gives

$$B(e, v) = \sum_{K \in \mathcal{P}} \int_K r(v - \mathcal{I}_X v) \, \mathrm{d}\boldsymbol{x} + \sum_{\gamma \in \partial \mathcal{P}} \int_\gamma R(v - \mathcal{I}_X v) \, \mathrm{d}s \quad \forall v \in V. \tag{2.11}$$

The identity (2.11) plays an important role, indirectly or directly, throughout *a posteriori* error analysis of finite element approximations. It may be used to derive an explicit *a posteriori* error estimate as follows. Applying the Cauchy–Schwarz inequality gives

$$B(e,v) \leq \sum_{K\in\mathcal{P}} \|r\|_{L_2(K)} \|v - \mathcal{I}_X v\|_{L_2(K)} + \sum_{\gamma\in\partial\mathcal{P}} \|R\|_{L_2(\gamma)} \|v - \mathcal{I}_X v\|_{L_2(\gamma)} . \tag{2.12}$$

Let $\widetilde{K}$ denote the subdomain consisting of elements sharing a common edge with element K as in equation (1.30),

$$\widetilde{K} = \operatorname{int}\left\{\bigcup K' \in \mathcal{P} : \overline{K}' \cap \overline{K} \text{ is nonempty}\right\}. \tag{2.13}$$

According to Theorem 1.7, there exists a constant C that is independent of v and h_K such that

$$\|v - \mathcal{I}_X v\|_{L_2(K)} \leq C h_K \|v\|_{H^1(\widetilde{K})} \tag{2.14}$$

and

$$\|v - \mathcal{I}_X v\|_{L_2(\partial K)} \leq C h_K^{1/2} \|v\|_{H^1(\widetilde{K})} \tag{2.15}$$

where h_K is the diameter of the element K. Inserting these estimates in inequality (2.12) and applying the Cauchy–Schwarz inequality leads to

$$B(e,v) \leq C \|v\|_{H^1(\Omega)} \left\{\sum_{K\in\mathcal{P}} h_K^2 \|r\|^2_{L_2(K)} + \sum_{\gamma\in\partial\mathcal{P}} h_K \|R\|^2_{L_2(\gamma)}\right\}^{1/2} . \tag{2.16}$$

Finally, thanks to the coercivity of the bilinear form over the global space V, it follows that $\|v\|_{H^1(\Omega)} \leq C |||v|||$, where $|||\cdot|||$ denotes the energy norm for the model problem. So, substituting e in place of v results in the *a posteriori* error estimate

$$|||e|||^2 \leq C \left\{\sum_{K\in\mathcal{P}} h_K^2 \|r\|^2_{L_2(K)} + \sum_{\gamma\in\partial\mathcal{P}} h_K \|R\|^2_{L_2(\gamma)}\right\}. \tag{2.17}$$

Apart from the constant C, all of the quantities on the right-hand side can be computed explicitly from the data and the finite element approximation. Typically, the terms on the right-hand side are regrouped as

$$|||e|||^2 \leq C \sum_{K\in\mathcal{P}} \left\{h_K^2 \|r\|^2_{L_2(K)} + h_K \|R\|^2_{L_2(\partial K)}\right\}, \tag{2.18}$$

and, in turn, this suggests defining a quantity η_K associated with element K to be

$$\eta_K^2 = h_K^2 \|r\|^2_{L_2(K)} + h_K \|R\|^2_{L_2(\partial K)} . \tag{2.19}$$

The *locally computed* quantity η_K is the contribution from element K to the bound for the *global error* $|||e|||$. It is important to recognize that the local quantity η_K need not necessarily provide a good estimate for the true local error $|||e|||_K$ in element K, due to the possibility of *pollution errors* arising from remote influences, such as singularities, that are insufficiently resolved. The important question of *local a posteriori error estimation* in the presence of pollution errors will be discussed in Chapter 8. Nevertheless, it is customary to assume that the local quantities η_K provide a sufficiently good guide to the actual errors $|||e|||_K$, for the local error indicators η_K to be employed for the purposes of driving an adaptive refinement algorithm.

2.3 EFFICIENCY OF ESTIMATOR

The *a posteriori* estimator implied by equation (2.18),

$$|||e|||^2 \leq C \sum_{K \in \mathcal{P}} \eta_K^2, \tag{2.20}$$

provides an upper bound on the discretization error, up to the (in general unknown) constant C. If the estimator is to be used as the basis of an adaptive refinement algorithm and, in particular, as a stopping criterion, then it is desirable for the estimator to be *efficient* in the sense that a constant C should exist, which does not depend on the mesh size, such that

$$\sum_{K \in \mathcal{P}} \eta_K^2 \leq C \, |||e|||^2 . \tag{2.21}$$

This type of bound is of practical importance since, in conjunction with the lower bound (2.18), it confirms that the rate of change of the estimator as the mesh size is reduced mirrors the behavior of the actual error. If no such estimate were possible without the multiplier C degrading as the mesh size is reduced, then the performance of the estimator would be pessimistic, and its use in driving an adaptive algorithm would result in meshes that were poorly designed and a tendency for the algorithm to fail to terminate until increasingly far beyond the point where the stopping criterion was actually met.

The problem of obtaining two-sided bounds for explicit estimators such as the one derived in Section 2.2 was addressed by Verfürth [111], who devised a technique that will be the subject of the following sections.

2.3.1 Bubble Functions

A key role will be played by certain locally supported, nonnegative functions that are commonly referred to as *bubble functions*. The two types of bubble function are *interior* bubble functions, supported on a single element, and *edge* bubble functions supported on a pair of elements. To begin with, it is convenient to define the bubble functions on the reference element, and to later construct the bubble functions on the physical element using the standard finite element approach based on the mapping from the reference element to the physical element.

The bubble functions on the triangular and quadrilateral reference elements are defined as follows.

Triangular Elements Let the triangular reference element be chosen as

$$\widehat{K} = \{(\widehat{x}, \widehat{y}) : 0 \leq \widehat{x} \leq 1; \quad 0 \leq \widehat{y} \leq 1 - \widehat{x}\} \tag{2.22}$$

and introduce the barycentric, or area, coordinates on the reference element defined by

$$\widehat{\lambda}_1 = \widehat{x}; \quad \widehat{\lambda}_2 = \widehat{y}; \quad \widehat{\lambda}_3 = 1 - \widehat{x} - \widehat{y}. \tag{2.23}$$

The interior bubble function $\widehat{\psi}$ is defined by

$$\widehat{\psi} = 27\widehat{\lambda}_1\widehat{\lambda}_2\widehat{\lambda}_3 \tag{2.24}$$

and the three edge bubble functions are given by

$$\widehat{\chi}_1 = 4\widehat{\lambda}_2\widehat{\lambda}_3; \quad \widehat{\chi}_2 = 4\widehat{\lambda}_1\widehat{\lambda}_3; \quad \widehat{\chi}_3 = 4\widehat{\lambda}_1\widehat{\lambda}_2. \tag{2.25}$$

The interior and edge bubble functions on a patch of triangular elements are shown in Figure 2.1.

Quadrilateral Elements The quadrilateral reference element is chosen to be the square

$$\widehat{K} = \{(\widehat{x}, \widehat{y}) : -1 \leq \widehat{x} \leq 1; \quad -1 \leq \widehat{y} \leq 1\}. \tag{2.26}$$

The interior bubble function $\widehat{\psi}$ is defined by

$$\widehat{\psi} = (1 - \widehat{x}^2)(1 - \widehat{y}^2) \tag{2.27}$$

and the four edge bubble functions are given by

$$\left.\begin{aligned} \widehat{\chi}_1 &= \tfrac{1}{2}(1 - \widehat{x}^2)(1 - \widehat{y}) \\ \widehat{\chi}_2 &= \tfrac{1}{2}(1 - \widehat{x})(1 - \widehat{y}^2) \\ \widehat{\chi}_3 &= \tfrac{1}{2}(1 - \widehat{x}^2)(1 + \widehat{y}) \\ \widehat{\chi}_4 &= \tfrac{1}{2}(1 + \widehat{x})(1 - \widehat{y}^2). \end{aligned}\right\} \tag{2.28}$$

Roughly speaking, the chief purpose of the interior bubble functions is to localize residuals to a single element. For example, if the interior residual r, defined in (2.4), is multiplied by the interior bubble function ψ_K for the element, then $v = r\psi_K$ vanishes on the element boundary. Consequently, v

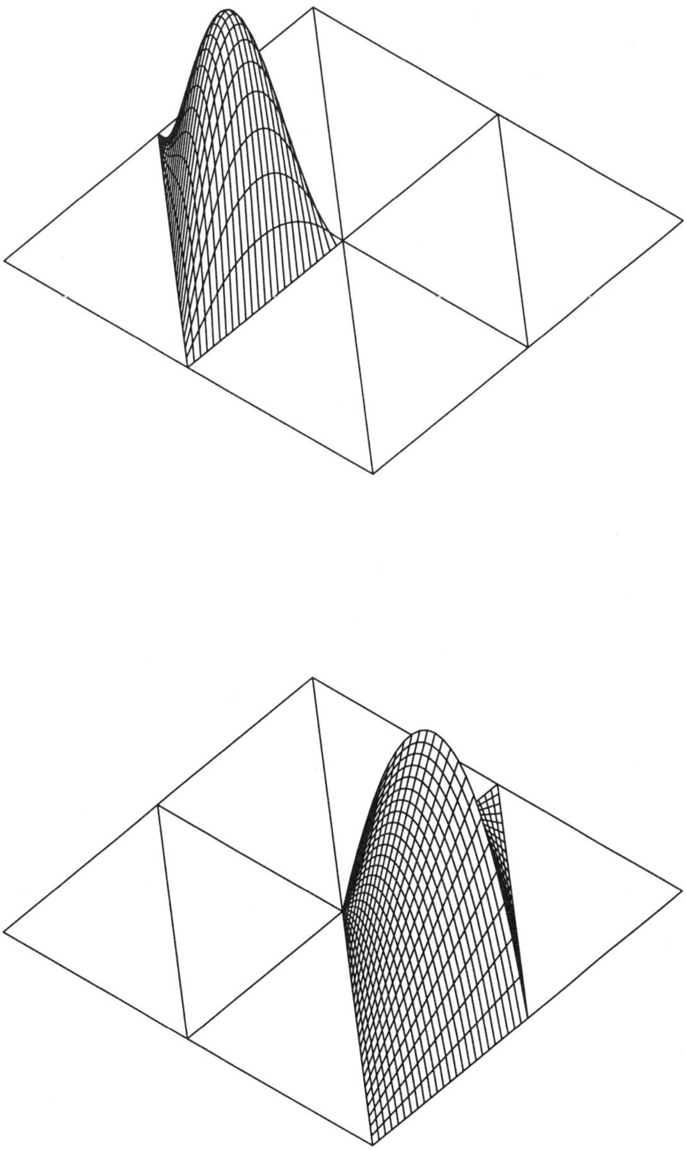

Fig. 2.1 Typical interior and edge bubble functions on a patch of triangular elements.

may be extended to the whole domain by assigning its value to be zero on the remaining elements. If the function r is sufficiently smooth, then the extended function v belongs to the space of admissible displacements V.

It will be useful to consider the effect of multiplication by the bubble function ψ_K on the norm of a residual. The following result shows that on the reference element, under certain additional assumptions, the norm is preserved up to a constant. This preliminary result concerning the reference element will then form the basis for an estimate on the physical element.

Lemma 2.1 *Let $\widehat{P} \subset H^1(\widehat{K})$ be a finite-dimensional subspace and let $\widehat{\psi}$ denote the interior bubble function over the reference element $\widehat{K}$. Then, for all $\widehat{v} \in \widehat{P}$, there exists a constant $\widehat{C}$ such that*

$$\widehat{C}^{-1} \|\widehat{v}\|^2_{L_2(\widehat{K})} \leq \int_{\widehat{K}} \widehat{\psi}\widehat{v}^2 \, \mathrm{d}\widehat{x} \leq \widehat{C} \|\widehat{v}\|^2_{L_2(\widehat{K})} \tag{2.29}$$

and

$$\widehat{C}^{-1} \|\widehat{v}\|_{L_2(\widehat{K})} \leq \left\|\widehat{\psi}\widehat{v}\right\|_{H^1(\widehat{K})} \leq \widehat{C} \|\widehat{v}\|_{L_2(\widehat{K})} \tag{2.30}$$

where the constant $\widehat{C}$ depends only on the reference element.

Proof. Thanks to the nonvanishing of the bubble functions on the interior of the element, it is easily seen that the mappings

$$\widehat{v} \mapsto \left\{\int_{\widehat{K}} \widehat{\psi}\widehat{v}^2 \, \mathrm{d}\widehat{x}\right\}^{1/2}$$

and

$$\widehat{v} \mapsto \left\|\widehat{\psi}\widehat{v}\right\|_{H^1(\widehat{K})}$$

both define norms on the finite-dimensional space $\widehat{P}$. The results then follow immediately from the fact that all norms on a finite-dimensional space are equivalent. ■

The next step is to construct bubble functions on the physical elements. For each element $K \in \mathcal{P}$ let $\mathcal{F}_K : \widehat{K} \mapsto K$ be the mapping described in Sections 1.3.4 and 1.3.5 and define the bubble functions on element K by

$$\psi_K = \widehat{\psi} \circ \mathcal{F}_K^{-1}; \quad \chi_\gamma = \widehat{\chi} \circ \mathcal{F}_K^{-1}. \tag{2.31}$$

If the partition $\mathcal{P}$ is regular, then, as in Sections 1.3.4 and 1.3.5, it follows that the mapping $\mathcal{F}_K$ satisfies

$$\left.\begin{aligned} \|D\mathcal{F}_K\|_{L_\infty(\widehat{K})} &\leq Ch_K \\ \|D\mathcal{F}_K^{-1}\|_{L_\infty(\widehat{K})} &\leq C\kappa_K/\rho_K \\ C\rho_K^2 \leq \|J_K\|_{L_\infty(\widehat{K})} &\leq Ch_K^2 \end{aligned}\right\} \tag{2.32}$$

where J_K and $D\mathcal{F}_K$ denote, respectively, the Jacobian and the Jacobian matrix of the transformation $\mathcal{F}_K$.

The finite-dimensional subspace P on the physical elements is constructed from the subspace $\widehat{P}$ on the reference element in the same fashion as the pull-back functions described earlier,

$$P = \left\{ \widehat{v} \circ \mathcal{F}_K^{-1} : \widehat{v} \in \widehat{P} \right\}. \tag{2.33}$$

The next result shows that the interior bubble function on the physical element has properties analogous to those described for the reference element in Lemma 2.1.

Theorem 2.2 *There exists a constant C such that for all $v \in P$*

$$C^{-1} \|v\|_{L_2(K)}^2 \le \int_K \psi v^2 \, \mathrm{d}\boldsymbol{x} \le C \|v\|_{L_2(K)}^2 \tag{2.34}$$

and

$$C^{-1} \|v\|_{L_2(K)} \le \|\psi v\|_{L_2(K)} + h_K |\psi v|_{H^1(K)} \le C \|v\|_{L_2(K)} \tag{2.35}$$

where the constant C is independent of v and h_K.

Proof. The result follows by mapping to the reference domain $\widehat{K}$, applying Lemma 2.1 and a scaling argument. ■

The role of the edge bubble function is to extend quantities defined on the element interfaces, such as jump residuals, to the pair of elements sharing the interface. The following result is analogous to Lemma 2.1, and it shows that the extension preserves norms, again up to a constant.

Lemma 2.3 *Let $\widehat{\gamma} \subset \partial\widehat{K}$ be an edge and let $\widehat{\chi}$ be the corresponding edge bubble function. Let $\widehat{P}(\widehat{\gamma})$ be a finite-dimensional space of functions defined on $\widehat{\gamma}$. Then there exists a constant $\widehat{C}$ such that for all $\widehat{v} \in \widehat{P}$,*

$$\widehat{C}^{-1} \|\widehat{v}\|_{L_2(\widehat{\gamma})}^2 \le \int_{\widehat{\gamma}} \widehat{\chi} \widehat{v}^2 \, \mathrm{d}\widehat{s} \le \widehat{C} \|\widehat{v}\|_{L_2(\widehat{\gamma})}^2 \tag{2.36}$$

and

$$\left\| \widehat{\mathcal{E}} \widehat{\chi} \widehat{v} \right\|_{H^1(\widehat{K})} \le \widehat{C} \|\widehat{v}\|_{L_2(\widehat{\gamma})}. \tag{2.37}$$

In both (2.36) and (2.37), the constants $\widehat{C}$ are independent of the function $\widehat{v}$.

Proof. The proof is based on the equivalence of norms on a finite-dimensional space. ■

This result has the following analogue on the physical element.

Theorem 2.4 *Let $\gamma \subset \partial K$ be an edge and let χ_γ be the corresponding edge bubble function. Let $P(\gamma)$ be the finite-dimensional space of functions defined on γ obtained by mapping $\widehat{P}(\widehat{\gamma}) \subset H^1(\widehat{\gamma})$. Then there exists a constant C such that*

$$C^{-1} \|v\|^2_{L_2(\gamma)} \leq \int_\gamma \chi_\gamma v^2 \, \mathrm{d}s \leq C \|v\|^2_{L_2(\gamma)} \tag{2.38}$$

and

$$h_K^{-1/2} \|\chi_\gamma v\|_{L_2(K)} + h_K^{1/2} |\chi_\gamma v|_{H^1(K)} \leq C \|v\|_{L_2(\gamma)} \tag{2.39}$$

where the constant C is independent of v and h_K.

Proof. The results follow at once from Lemma 2.3 using a scaling argument. ■

2.3.2 Bounds on the Residuals

The proof that the estimator provides two-sided bounds on the true error makes use of the residual equation (2.9) in conjunction with special choices of the function v.

The first task is to bound the quantity $\|r\|_{L_2(K)}$ in terms of the norm of the true error evaluated over a neighborhood of the element K. Let $\overline{r}$ be a piecewise discontinuous approximation to the interior residual r on element K from a suitable finite-dimensional subspace to be discussed later. For instance, if the original Galerkin approximation consists of piecewise affine functions over triangular elements, then one would use a piecewise affine approximation $\overline{r}$ to r on each element. Observe that there is no obligation for the approximation to be continuous across element boundaries. Applying Theorem 2.2 gives

$$\|\overline{r}\|^2_{L_2(K)} \leq C \int_K \psi_K \overline{r}^2 \, \mathrm{d}\boldsymbol{x}. \tag{2.40}$$

The function $v = \overline{r}\psi_K$ vanishes on the boundary of element K, and therefore, v may be extended to the rest of the domain as a continuous function by defining its value outside the element to be zero. The resulting function, again denoted by v, belongs to the space V since v is nonzero on element K and vanishes on the boundary ∂K. Thus, inserting v into the residual equation (2.9) yields

$$B(e, \overline{r}\psi_K) = \int_K \psi_K r \overline{r} \, \mathrm{d}\boldsymbol{x} \tag{2.41}$$

and so

$$\int_K \psi_K \overline{r}^2 \, \mathrm{d}\boldsymbol{x} = \int_K \psi_K \overline{r}(\overline{r} - r) \, \mathrm{d}\boldsymbol{x} + B(e, \overline{r}\psi_K). \tag{2.42}$$

The first term on the right-hand side may be bounded with the aid of the Cauchy–Schwarz inequality,

$$\int_K \psi_K \overline{r}(\overline{r} - r) \, \mathrm{d}\boldsymbol{x} \leq \|\psi_K \overline{r}\|_{L_2(K)} \|\overline{r} - r\|_{L_2(K)} \tag{2.43}$$

and then, applying the second part of Theorem 2.2, we obtain

$$\|\psi_K\overline{r}\|_{L_2(K)} \leq C\,\|\overline{r}\|_{L_2(K)}\,. \tag{2.44}$$

The second term is also dealt with by applying the Cauchy–Schwarz inequality

$$B(e,\overline{r}\psi_K) \leq \|\!|e|\!\|_K\,\|\psi_K\overline{r}\|_{H^1(K)}\,, \tag{2.45}$$

and then, again applying the second part of Theorem 2.2, we obtain

$$\|\psi_K\overline{r}\|_{H^1(K)} \leq Ch_K^{-1}\,\|\overline{r}\|_{L_2(K)}\,. \tag{2.46}$$

Therefore, inserting these estimates into equation (2.42) gives

$$\int_K \psi_K\overline{r}^2\,\mathrm{d}\boldsymbol{x} \leq C\,\|\overline{r}\|_{L_2(K)}\,\|\overline{r}-r\|_{L_2(K)} + Ch_K^{-1}\,\|\!|e|\!\|_K\,\|\overline{r}\|_{L_2(K)} \tag{2.47}$$

and recalling (2.40) leads to the bound

$$\|\overline{r}\|_{L_2(K)} \leq C\left\{h_K^{-1}\,\|\!|e|\!\|_K + \|\overline{r}-r\|_{L_2(K)}\right\}. \tag{2.48}$$

The desired bound on the actual residual follows from this estimate and an application of the triangle inequality,

$$\|r\|_{L_2(K)} \leq C\left\{h_K^{-1}\,\|\!|e|\!\|_K + \|\overline{r}-r\|_{L_2(K)}\right\}. \tag{2.49}$$

It remains to estimate the boundary residual $\|R\|_{L_2(\gamma)}$ in terms of the norm of the true error evaluated over the pair of elements sharing the edge γ. The argument proceeds in an analogous fashion to the treatment of the interior residual. Let $\overline{R}$ be an approximation to the boundary residual or jumps from a suitable finite-dimensional space. For example, if the original Galerkin approximation was constructed using piecewise affine functions on triangles, then one would select $\overline{R}$ to be the average value of R on the edge γ. As before, there is obligation for the approximation to be continuous between the differing edges. The first part of Theorem 2.4 shows that

$$\|\overline{R}\|^2_{L_2(\gamma)} \leq C\int_\gamma \chi_\gamma\overline{R}^2\,\mathrm{d}s. \tag{2.50}$$

Let $\widetilde{\gamma}$ denote the subdomain of Ω consisting of the union of the side γ and the pair of elements, K and K' say, sharing the common side γ. The function $v = \overline{R}\chi_\gamma$ vanishes on the boundary of the subdomain $\widetilde{\gamma}$ and is continuous. Extending v by zero to the whole of the domain Ω gives a function v from the space V. The residual equation (2.9), with this choice of v, yields

$$B(e,\overline{R}\chi_\gamma) = \int_{\widetilde{\gamma}} \chi_\gamma r\overline{R}\,\mathrm{d}\boldsymbol{x} + \int_\gamma \chi_\gamma R\overline{R}\,\mathrm{d}s, \tag{2.51}$$

and therefore

$$\int_\gamma \chi_\gamma \overline{R}^2\,\mathrm{d}s = \int_\gamma \chi_\gamma \overline{R}(\overline{R}-R)\,\mathrm{d}s + B(e,\chi_\gamma\overline{R}) - \int_{\widetilde{\gamma}} \chi_\gamma r\overline{R}\,\mathrm{d}s. \tag{2.52}$$

Each of these terms is dealt with using Theorem 2.4 and the Cauchy–Schwarz inequality. The first term is estimated by

$$\begin{aligned}\int_\gamma \chi_\gamma \overline{R}(\overline{R}-R)\,\mathrm{d}s &\leq \left\|\chi_\gamma\overline{R}\right\|_{L_2(\gamma)} \left\|\chi_\gamma(\overline{R}-R)\right\|_{L_2(\gamma)} \\ &\leq C\left\|\overline{R}\right\|_{L_2(\gamma)} \left\|\overline{R}-R\right\|_{L_2(\gamma)},\end{aligned}$$

while the second term is dealt with by arguing

$$B(e,\chi_\gamma\overline{R}) \leq C\,|\!|\!|e|\!|\!|_{\widetilde{\gamma}} \left\|\chi_\gamma\overline{R}\right\|_{H^1(\widetilde{\gamma})} \leq Ch_\gamma^{-1/2}\,|\!|\!|e|\!|\!|_{\widetilde{\gamma}} \left\|\overline{R}\right\|_{L_2(\gamma)}, \tag{2.53}$$

and the final term is bounded by

$$\int_{\widetilde{\gamma}} \chi_\gamma r\overline{R}\,\mathrm{d}s \leq \|r\|_{L_2(\widetilde{\gamma})} \left\|\chi_\gamma\overline{R}\right\|_{L_2(\widetilde{\gamma})} \leq Ch_\gamma^{1/2}\,\|r\|_{L_2(\widetilde{\gamma})} \left\|\overline{R}\right\|_{L_2(\gamma)}. \tag{2.54}$$

As a consequence of these estimates and the bound (2.50), we conclude that

$$\left\|\overline{R}\right\|_{L_2(\gamma)} \leq C\left\{\left\|\overline{R}-R\right\|_{L_2(\gamma)} + h_\gamma^{-1/2}\,|\!|\!|e|\!|\!|_{\widetilde{\gamma}} + h_\gamma^{1/2}\,\|r\|_{L_2(\widetilde{\gamma})}\right\}. \tag{2.55}$$

In turn, this leads to the desired estimate for the residual $\|R\|_{L_2(\gamma)}$ from the triangle inequality and applying the estimate (2.49) for the interior residual in terms of the true error, giving

$$\|R\|_{L_2(\gamma)} \leq C\left\{h_\gamma^{-1/2}\,|\!|\!|e|\!|\!|_{\widetilde{\gamma}} + h_\gamma^{1/2}\,\|\overline{r}-r\|_{L_2(\widetilde{\gamma})} + \left\|\overline{R}-R\right\|_{L_2(\gamma)}\right\}. \tag{2.56}$$

Summarizing these developments, we have proved:

Theorem 2.5 *Let r and R denote the interior and boundary residuals associated with the finite element approximation constructed from the subspace X. Suppose that $\overline{r}$ and $\overline{R}$ are piecewise continuous approximations to the interior and boundary residuals constructed from finite-dimensional subspaces. Then,*

$$\|r\|_{L_2(K)} \leq C\left\{h_K^{-1}\,|\!|\!|e|\!|\!|_K + \|\overline{r}-r\|_{L_2(K)}\right\} \tag{2.57}$$

and

$$\|R\|_{L_2(\gamma)} \leq C\left\{h_\gamma^{-1/2}\,|\!|\!|e|\!|\!|_{\widetilde{\gamma}} + h_\gamma^{1/2}\,\|\overline{r}-r\|_{L_2(\widetilde{\gamma})} + \left\|\overline{R}-R\right\|_{L_2(\gamma)}\right\}. \tag{2.58}$$

where C is a positive depending only on the shape regularity of the elements and the selection of the finite-dimensional subspaces used to approximate the interior and boundary residuals.

2.3.3 Proof of Two-Sided Bounds on the Error

The bounds on the residuals,

$$\|r\|_{L_2(K)} \leq C\left\{h_K^{-1} \|e\|_K + \|\overline{r} - r\|_{L_2(K)}\right\} \tag{2.59}$$

and

$$\|R\|_{L_2(\gamma)} \leq C\left\{h_\gamma^{-1/2} \|e\|_{\widetilde{\gamma}} + h_\gamma^{1/2} \|\overline{r} - r\|_{L_2(\widetilde{\gamma})} + \left\|\overline{R} - R\right\|_{L_2(\gamma)}\right\}, \tag{2.60}$$

given in Theorem 2.5 are employed in the analysis of the efficiency of the error estimator as follows. First, it is necessary to select the finite-dimensional subspaces with which to approximate the interior and boundary residuals. The interior residual r, restricted to a single element, is given by

$$r = f + \Delta u_X - cu_X. \tag{2.61}$$

Now, the finite element approximation u_X is constructed from a finite-dimensional subspace composed of (pull-back) polynomials. Thus, the approximation $\overline{r}$ to the residual on element K can be constructed from the finite-dimensional subspace spanned by functions of the form $v_X|_K$ and $\Delta v_X|_K$, where v_X belongs to the finite element subspace X. Equally well, the finite-dimensional subspace used to approximate the edge residual can be constructed in a similar implicit fashion. As a consequence, the perturbation term $r - \overline{r}$ reduces to the form $f - \overline{f}$. The perturbation term $R - \overline{R}$ for the edge residual actually vanishes on the interior edges and reduces to $g - \overline{g}$ on the exterior Neumann boundary. Therefore,

$$\|r\|_{L_2(K)} \leq C\left\{h_K^{-1} \|e\|_K + \left\|f - \overline{f}\right\|_{L_2(K)}\right\} \tag{2.62}$$

and

$$\|R\|_{L_2(\gamma)} \leq C\left\{h_\gamma^{-1/2} \|e\|_{\widetilde{\gamma}} + h_\gamma^{1/2} \left\|f - \overline{f}\right\|_{L_2(\widetilde{\gamma})} + \|g - \overline{g}\|_{L_2(\gamma\cap\Gamma_N)}\right\}. \tag{2.63}$$

The local error indicator η_K in (2.19) associated with the element K can then be bounded by

$$\eta_K^2 \leq C\left\{\|e\|_{\widetilde{K}}^2 + h_K^2 \left\|f - \overline{f}\right\|_{L_2(\widetilde{K})}^2 + \sum_{\gamma\subset\partial K\cap\Gamma_N} h_K \|g - \overline{g}\|_{L_2(\gamma)}^2\right\} \tag{2.64}$$

where the constant C depends only on the regularity κ_K of the element.

The estimate shows that the error indicator is local in a certain sense, since the terms on the right-hand bound involve only contributions from the actual element and its immediate neighbors. Summarizing these developments, we have proved the following:

Theorem 2.6 *Let η_K denote the local error indicator*

$$\eta_K^2 = h_K^2 \left\|r\right\|_{L_2(K)}^2 + \frac{1}{2} h_K \left\|R\right\|_{L_2(\partial K)}^2 \tag{2.65}$$

where r and R are the interior and boundary residuals. Then there exists a constant C depending only on the regularity of the elements such that

$$\frac{1}{C} \left\|\!\left| e \right|\!\right\|^2 \le \sum_{K \in \mathcal{P}} \eta_K^2 \tag{2.66}$$

and

$$\eta^2 \le C \left\{ \left\|\!\left| e \right|\!\right\|^2 + \sum_{K \in \mathcal{P}} h_K^2 \left\| f - \overline{f} \right\|_{L_2(K)}^2 + \sum_{\gamma \subset \Gamma_N} h_K \left\| g - \overline{g} \right\|_{L_2(\gamma)}^2 \right\}. \tag{2.67}$$

Moreover, the local bound

$$\eta_K^2 \le C \left\{ \left\|\!\left| e \right|\!\right\|_{\widetilde{K}}^2 + h_K^2 \left\| f - \overline{f} \right\|_{L_2(\widetilde{K})}^2 + \sum_{\gamma \subset \partial K \cap \Gamma_N} h_K \left\| g - \overline{g} \right\|_{L_2(\gamma)}^2 \right\} \tag{2.68}$$

also holds.

Proof. Follows at once from previous arguments. ■

Generally, the data f and g may be well-approximated by the piecewise continuous functions $\overline{f}$ and $\overline{g}$, meaning that the terms involving the differences $f - \overline{f}$ and $g - \overline{g}$ will be relatively small compared to the true error. In this sense, the estimator obtained by summing the local indicators provides an equivalent measure of the actual error in the energy norm. The local bound (2.68) is of importance for the design of adaptive algorithms since it shows that the estimator gives some indication of the distribution of the true error in the sense that if the indicator η_K is large on a particular element, then the true error must also be large in a neighborhood of the element. Therefore, refining those elements where the indicator is large will target refinements towards the regions where the true error is large. Of course, the converse statement is, in general, false owing to the possibility of pollution effects.

2.4 A SIMPLE EXPLICIT LEAST SQUARES ERROR ESTIMATOR

The duality argument due to Aubin and Nitsche, as described, for instance in Ciarlet [50], plays an important role in the derivation of *a priori* error estimates in norms other than the energy norm. Likewise, the technique may also be used for the purpose of deriving *a posteriori* error estimators.

The starting point for the application of the technique is the consideration of the adjoint of the original model problem:

$$\Phi_F \in V : B(v, \Phi_F) = (F, v) \quad \forall v \in V \tag{2.69}$$

where $F \in L_2(\Omega)$ is any sufficiently smooth load. It is assumed that this problem is *regular* in the sense that the solution Φ_F satisfies $\Phi_F \in H^2(\Omega) \cap V$ and that there exists a constant C such that

$$\|\Phi_F\|_{H^2(\Omega)} \leq C \|F\|_{L_2(\Omega)} . \tag{2.70}$$

This assumption is known to hold, in particular, if the domain Ω is convex [50]. The specific choice of $F = e$, the true error, then gives

$$\|e\|^2_{L_2(\Omega)} = B(e, \Phi_e). \tag{2.71}$$

The residual equation (2.11) plays a lead role similarly to the derivation of the energy norm estimator. Selecting $v = \Phi_e$ in (2.11) reveals that

$$B(e, \Phi_e) = \sum_{K \in \mathcal{P}} \int_K r(\Phi_e - \mathcal{I}_X \Phi_e)\, \mathrm{d}\boldsymbol{x} + \sum_{\gamma \in \partial\mathcal{P}} \int_\gamma R(\Phi_e - \mathcal{I}_X \Phi_e)\, \mathrm{d}s \tag{2.72}$$

and then, as before, the Cauchy–Schwarz inequality gives

$$\begin{aligned} B(e, v) \;\leq\; & \sum_{K \in \mathcal{P}} \|r\|_{L_2(K)} \|\Phi_e - \mathcal{I}_X \Phi_e\|_{L_2(K)} \\ + \; & \sum_{\gamma \in \partial\mathcal{P}} \|R\|_{L_2(\gamma)} \|\Phi_e - \mathcal{I}_X \Phi_e\|_{L_2(\gamma)} . \end{aligned} \tag{2.73}$$

Slightly different approximation theoretic results are employed compared to those used in (2.14) and (2.15) to derive the estimator for the error measured in the energy norm. Instead, the additional regularity of the solution Φ_e allows an extra power of the mesh size to be extracted through the use of the following estimates arising from Theorem 1.7: There exists a constant C that is independent of v and h_K such that

$$\|\Phi_e - \mathcal{I}_X \Phi_e\|_{L_2(K)} \leq C h_K^2 \|\Phi_e\|_{H^2(\widetilde{K})} \tag{2.74}$$

and

$$\|\Phi_e - \mathcal{I}_X \Phi_e\|_{L_2(\partial K)} \leq C h_K^{3/2} \|\Phi_e\|_{H^2(\widetilde{K})} . \tag{2.75}$$

Inserting these estimates into (2.73) and applying the Cauchy–Schwarz inequality leads to

$$\|e\|^2_{L_2(\Omega)} \leq C \|\Phi_e\|_{H^2(\Omega)} \left\{ \sum_{K \in \mathcal{P}} h_K^4 \|r\|^2_{L_2(K)} + \sum_{\gamma \in \partial\mathcal{P}} h_K^3 \|R\|^2_{L_2(\gamma)} \right\}^{1/2} \tag{2.76}$$

and so, with the aid of the bound (2.70), there follows

$$\|e\|_{L_2(\Omega)} \le C \left\{ \sum_{K\in\mathcal{P}} h_K^4 \|r\|^2_{L_2(K)} + \sum_{\gamma\in\partial\mathcal{P}} h_K^3 \|R\|^2_{L_2(\gamma)} \right\}^{1/2}. \tag{2.77}$$

A rearrangement of the terms on the right-hand side gives an estimator similar to the one derived before, with the only difference being a higher-order scaling by the mesh size reflecting the expectation of a higher rate of convergence of the approximation in the L_2 norm,

$$\|e\|^2_{L_2(\Omega)} \le C \sum_{K\in\mathcal{P}} \left\{ h_K^4 \|r\|^2_{L_2(K)} + h_K^3 \|R\|^2_{L_2(\partial K)} \right\}. \tag{2.78}$$

This result is recorded for future reference.

Theorem 2.7 *Suppose that the domain Ω is convex and that $\Gamma_D = \partial\Omega$. Let $\eta_{L_2(K)}$ denote the local error indicator*

$$\eta^2_{L_2(K)} = h_K^4 \|r\|^2_{L_2(K)} + h_K^3 \|R\|^2_{L_2(\partial K)} \tag{2.79}$$

where r and R are the interior and boundary residuals. Then there exists a constant C depending on the domain Ω and the shape of the elements such that

$$\|e\|^2_{L_2(\Omega)} \le C \sum_{K\in\mathcal{P}} \eta^2_{L_2(K)}. \tag{2.80}$$

Proof. It is known [50] that the adjoint problem (2.69) satisfies the regularity assumption (2.70) whenever the domain Ω is convex. The proof then follows immediately from the above arguments. ■

The assumption on the convexity of the domain is needed to ensure that the regularity assumption is satisfied. Unfortunately, the assumption is rather restrictive and excludes the important case of re-entrant corners. While it is possible to weaken the assumption and work with a milder regularity property, this would result in a smaller exponent of the mesh size in the estimates that would depend on the geometry and boundary conditions. The next section is concerned with estimators for the pointwise errors. In contrast to the results presented in this section, the analysis will be valid for domains with re-entrant corners and two-sided bounds will be obtained.

2.5 ESTIMATES FOR THE POINTWISE ERROR

The purpose of this section is to derive *a posteriori* estimates for the pointwise error $\|e\|_{L_\infty(\Omega)}$, similar to the explicit estimates obtained for the error measured in the energy norm. For ease of exposition, it will be assumed throughout that the finite element approximation $u_X \in X$ of the pure Dirichlet problem

$$-\Delta u = f \text{ in } \Omega; \quad u = 0 \text{ on } \partial\Omega \tag{2.81}$$

where $f \in L_\infty(\Omega)$, is constructed using piecewise affine functions on a regular partition of the domain Ω consisting of triangles. In addition, the fairly mild assumption is made that the size h_{max} (respectively h_{min}) of the largest (respectively smallest) element satisfies

$$h_{min} \geq C h_{max}^\gamma \tag{2.82}$$

where $C > 0$ and $\gamma \geq 1$ are fixed constants. It is assumed that this condition holds for all of the meshes that are generated in the sequence of adaptive refinements with the same values of the constants C and γ. In particular, while it is still possible for the mesh to be highly locally refined, situations where elements present in the initial mesh remain unrefined throughout the whole sequence of adaptive steps are disallowed. The same assumption holds trivially in the context of a single mesh; however, all of the constants that arise in the analysis will depend on the values of C and γ.

It is also worth recording that, in all but trivial cases when the true solution u is piecewise affine, there will be at least one element $K \in \mathcal{P}$ for which

$$\inf_{v_X \in X} \|u - v_X\|_{L_\infty(K)} \geq C(u) h_K^2. \tag{2.83}$$

However, in view of the possibility that the true solution u may belong to the finite element space X, it is necessary to make a formal assumption that (2.83) holds.

2.5.1 Regularized Point Load

Let $\boldsymbol{x}_0 \in \Omega$ and $\rho_0 > 0$ be quantities to be chosen later. A regularized point load, or Dirac mass, $\delta \in C_0^\infty(\Omega)$ is constructed so that

$$\int_\Omega \delta(\boldsymbol{x})\, \mathrm{d}\boldsymbol{x} = 1, \tag{2.84}$$

and such that there exists a positive constant C, independent of ρ_0, for which

$$0 \leq \delta(\boldsymbol{x}) \leq C\rho_0^{-2} \quad \forall \boldsymbol{x} \in \Omega. \tag{2.85}$$

In addition, δ is assumed to be locally supported, $\operatorname{supp}\delta \subset \mathcal{B}_0$, over the domain $\mathcal{B}_0$ consisting of the points in Ω that are contained in a ball of radius $\rho_0/2$ centered around the point $\boldsymbol{x}_0$,

$$\mathcal{B}_0 = \{\boldsymbol{x} \in \Omega : \operatorname{dist}(\boldsymbol{x}, \boldsymbol{x}_0) \leq \rho_0/2\}. \tag{2.86}$$

The first result relates the pointwise error to a weighted average of the error in the neighborhood of where the pointwise error is largest.

Lemma 2.8 *Let a point $\boldsymbol{x}_0 \in \Omega$ be selected such that $\|e\|_{L_\infty(\Omega)} = |e(\boldsymbol{x}_0)|$. Then, under the previous assumptions, there exist constants C and h^* such that for $h_{max} \leq h^*$ sufficiently small,*

$$\|e\|_{L_\infty(\Omega)} \leq C\,|(\delta, e)| \tag{2.87}$$

where C is a positive constant depending on h^*.

Proof. Introduce the domain $\widetilde{\mathcal{B}}_0$ consisting of the elements that intersect the support $\mathcal{B}_0$ of the regularized Dirac mass,

$$\widetilde{\mathcal{B}}_0 = \operatorname{int}\left\{\bigcup \overline{K}, K \in \mathcal{P} : K \cap \mathcal{B}_0 \text{ is nonempty}\right\}.$$

The assumption on the regularity of the mesh means that all of elements contained in the extended domain $\widetilde{\mathcal{B}}_0$ have diameters of order h_0, where h_0 is the diameter of an element containing the point $\boldsymbol{x}_0$. The parameter ρ_0 is then taken to be $\rho_0 = h_0^\beta$, where $\beta > 2$ is a constant to be determined. Finally, a point $\boldsymbol{x}_1 \in \mathcal{B}_0$ may be chosen such that

$$e(\boldsymbol{x}_1) = (\delta, e).$$

The true solution u is known [69] to be Hölder continuous with exponent $0 < \alpha \leq 1$, so that, in particular, u is continuous. Now, observe that

$$I_X e(\boldsymbol{x}) = I_X u(\boldsymbol{x}) - u_X(\boldsymbol{x})$$

and so

$$e(\boldsymbol{x}) = u(\boldsymbol{x}) + I_X e(\boldsymbol{x}) - I_X u(\boldsymbol{x}).$$

Hence, thanks to the triangle inequality, we obtain

$$\begin{aligned} |e(\boldsymbol{x}_0) - e(\boldsymbol{x}_1)| \;\leq\; & |u(\boldsymbol{x}_0) - u(\boldsymbol{x}_1)| + |I_X e(\boldsymbol{x}_0) - I_X e(\boldsymbol{x}_1)| \\ & + |I_X u(\boldsymbol{x}_0) - I_X u(\boldsymbol{x}_1)| \end{aligned} \tag{2.88}$$

where $I_X u \in X$ denotes the piecewise linear interpolant to u.

The first term may be bounded with the aid of the Hölder continuity of the true solution as follows

$$|u(\boldsymbol{x}_0) - u(\boldsymbol{x}_1)| \leq C|\boldsymbol{x}_0 - \boldsymbol{x}_1|^\alpha \leq C\rho_0^\alpha \leq Ch_0^{\alpha\beta}.$$

Applying assumptions (2.82) and (2.83) it follows that

$$Ch_0^{2\gamma} \leq Ch_{max}^{2\gamma} \leq Ch_{min}^2 \leq Ch_K^2 \leq C\,\|u - u_X\|_{L_\infty(K)} \leq C\,\|e\|_{L_\infty(\Omega)}\,.$$

By choosing $\beta > 2\gamma/\alpha \geq 2$, it follows that

$$h_0^{\alpha\beta} \leq Ch_0^{2\gamma+\nu} \leq Ch_0^\nu\,\|e\|_{L_\infty(\Omega)}$$

where $\nu = \alpha(\beta - 2\gamma/\alpha) > 0$, and so

$$|u(\boldsymbol{x}_0) - u(\boldsymbol{x}_1)| \leq Ch_0^{\nu} \|e\|_{L_\infty(\Omega)} .$$

The second term in (2.88) is bounded by arguing that

$$|I_X e(\boldsymbol{x}_0) - I_X e(\boldsymbol{x}_1)| \leq C\rho_0 \|\boldsymbol{\nabla} I_X e\|_{L_\infty(\widetilde{\mathcal{B}}_0)}$$

and then by an inverse estimate (see Theorem 1.3),

$$\|\boldsymbol{\nabla} I_X e\|_{L_\infty(\widetilde{\mathcal{B}}_0)} \leq Ch_0^{-1} \|I_X e\|_{L_\infty(\widetilde{\mathcal{B}}_0)} \leq Ch_0^{-1} \|e\|_{L_\infty(\Omega)} ,$$

so that

$$|I_X e(\boldsymbol{x}_0) - I_X e(\boldsymbol{x}_1)| \leq Ch_0^{\beta-1} \|e\|_{L_\infty(\Omega)} .$$

The final term in (2.88) is estimated by observing that

$$|I_X u(\boldsymbol{x}_0) - I_X u(\boldsymbol{x}_1)| \leq C\rho_0 \|\boldsymbol{\nabla} I_X u\|_{L_\infty(\widetilde{\mathcal{B}}_0)} .$$

Then, letting $\overline{u}_X$ denote the average of the values of the finite element approximation at the vertices $\{\boldsymbol{x}_k\}$ of the elements contained in the set $\widetilde{\mathcal{B}}_0$, and again with the aid of an inverse estimate, we obtain

$$\begin{aligned} \|\boldsymbol{\nabla} I_X u\|_{L_\infty(\widetilde{\mathcal{B}}_0)} &= \|\boldsymbol{\nabla} I_X (u - \overline{u}_X)\|_{L_\infty(\widetilde{\mathcal{B}}_0)} \\ &\leq Ch_0^{-1} \|I_X (u - \overline{u}_X)\|_{L_\infty(\widetilde{\mathcal{B}}_0)} \\ &\leq Ch_0^{-1} \max_k |e(\boldsymbol{x}_k)| \\ &\leq Ch_0^{-1} \|e\|_{L_\infty(\Omega)} . \end{aligned}$$

Hence,

$$|I_X u(\boldsymbol{x}_0) - I_X u(\boldsymbol{x}_1)| \leq Ch_0^{\beta-1} \|e\|_{L_\infty(\Omega)} .$$

Inserting these estimates into (2.88), it follows that

$$|e(\boldsymbol{x}_0) - e(\boldsymbol{x}_1)| \leq Ch_0^{\min(\nu,\beta-1)} \|e\|_{L_\infty(\Omega)} .$$

Applying the triangle inequality and recalling the property leading to the definition of the point $\boldsymbol{x}_1$ leads to the conclusion

$$(1 - Ch_0^{\min(\nu,\beta-1)}) \|e\|_{L_\infty(\Omega)} \leq |(\delta, e)|;$$

therefore for $h_0 \leq h_{max} \leq h^*$ sufficiently small we obtain

$$\|e\|_{L_\infty(\Omega)} \leq C|(\delta, e)|,$$

and the result then follows as claimed. ■

2.5.2 Regularized Green's Function

A regularized Green's function $G \in H_0^1(\Omega)$ is associated with the regularized Dirac mass δ through the definition

$$B(v, G) = (\delta, v) \quad \forall v \in H_0^1(\Omega). \tag{2.89}$$

It is shown in Nochetto [85, Theorem 3.1] that, for each finite $p > 1$, the second derivatives of the regularized Green's function may be bounded by

$$|G|_{W^{2,p}(\Omega)} \leq \frac{C\rho_0^{-4(p-1)}}{(p-1)^2} \tag{2.90}$$

where C is a positive constant that is independent of ρ_0 and p. Thus, selecting $p = 1 + |\log \rho_0|^{-1}$ and recalling $\rho_0 = h_0^\beta$ leads to the conclusion

$$|G|_{W^{2,1}(\Omega)} \leq C|\log h_{max}|^2. \tag{2.91}$$

The derivation of the *a posteriori* bound begins with choosing $v = e$ in (2.89), giving

$$(\delta, e) = B(e, G). \tag{2.92}$$

Arguing as in the derivation of the residual equation (2.11), one readily concludes that

$$B(e, G) = \sum_{K \in \mathcal{P}} \int_K r(G - I_X G)\, \mathrm{d}x + \sum_{\gamma \in \partial\mathcal{P}} \int_\gamma R(G - I_X G)\, \mathrm{d}s \tag{2.93}$$

where $I_X G \in X$ is the interpolant of G as in Theorem 1.5. The terms are bounded using Hölder's inequality leading to the conclusion

$$\begin{aligned} |B(e, G)| \ &\leq \ \sum_{K \in \mathcal{P}} \|r\|_{L_\infty(K)} \|G - I_X G\|_{L_1(K)} \\ &\quad + \sum_{\gamma \in \partial\mathcal{P}} \|R\|_{L_\infty(\gamma)} \|G - I_X G\|_{L_1(K)} \, . \end{aligned} \tag{2.94}$$

According to Theorem 1.5, there exists a constant C such that

$$\|G - I_X G\|_{L_1(K)} \leq C h_K^2 \, |G|_{W^{2,1}(\widetilde{K})} \tag{2.95}$$

and

$$\|G - I_X G\|_{L_1(\gamma)} \leq C h_\gamma \, |G|_{W^{2,1}(\widetilde{\gamma})} \, . \tag{2.96}$$

Inserting these estimates into (2.94) and applying (2.88) leads to the bound

$$|B(e, G)| \leq C\eta_\infty \, |G|_{W^{2,1}(\Omega)} \leq C|\log h_{max}|^2 \eta_\infty$$

where

$$\eta_\infty = \max_{K \in \mathcal{P}} h_K^2 \, \|r\|_{L_\infty(K)} + \max_{\gamma \in \partial\mathcal{P}} h_\gamma \, \|R\|_{L_\infty(\gamma)} \, . \tag{2.97}$$

Combining this with Lemma 2.8 gives

$$\|e\|_{L_\infty(\Omega)} \leq C|\log h_{max}|^2 \eta_\infty. \tag{2.98}$$

For the purposes of practical computation, the logarithmic factor appearing in the bound (2.98) may be regarded as being constant and the quantity η_∞ is used as an estimator for the pointwise error.

2.5.3 Two-Sided Bounds on the Pointwise Error

The purpose of this section is to show that the estimator η_∞ is efficient in the sense that η_∞ is, up to a constant, bounded above by the true error.

The discussion is similar to the one used to obtain two-sided bounds for the error measured in the energy norm. In particular, use is made of the bubble functions defined in Section 2.3.1 along with piecewise constant approximations $\overline{r}$ and $\overline{R}$ to the actual residuals.

Let K' be any element and define $v \in H_0^1(\Omega)$ to be

$$v = \sum_{K \subset \widetilde{K}'} \alpha_K \psi_K + \sum_{\gamma \subset \partial K'} \beta_\gamma \psi_\gamma \tag{2.99}$$

where ψ_K is the interior bubble function on element K and ψ_γ is the bubble function on edge γ. The parameters $\{\alpha_K\}$ and $\{\beta_\gamma\}$ are determined by the conditions

$$\int_K v \operatorname{sgn}(\overline{r})\, d\boldsymbol{x} = h_K^2 \quad \forall K \subset \widetilde{K}' \tag{2.100}$$

with

$$\int_\gamma v \operatorname{sgn}(\overline{R})\, ds = \begin{cases} h_\gamma & \text{if } \gamma \subset \partial K' \backslash \partial\Omega \\ 0 & \text{if } \gamma \subset \partial\Omega. \end{cases} \tag{2.101}$$

The latter condition determines β_γ directly, since, thanks to ψ_K vanishing on the element boundaries, we have

$$\int_\gamma v \operatorname{sgn}(\overline{R})\, ds = \beta_\gamma \int_\gamma \psi_\gamma \operatorname{sgn}(\overline{R})\, ds. \tag{2.102}$$

Consequently, if γ is an interior edge, we obtain

$$\beta_\gamma^{-1} = \frac{1}{h_\gamma} \int_\gamma \psi_\gamma \operatorname{sgn}(\overline{R})\, ds. \tag{2.103}$$

Evidently, the expression on the right-hand side is related to the average value of the bubble function on the edge and it therefore follows that β_γ is bounded above and below by positive constants independent of h_γ. If γ lies on the exterior boundary, then β_γ vanishes.

The coefficients $\{\alpha_K\}$ are determined by the condition (2.100)

$$h_K^2 = \int_K v \operatorname{sgn}(\overline{r})\, d\boldsymbol{x} = \alpha_K \int_K \psi_K \operatorname{sgn}(\overline{r})\, d\boldsymbol{x} + \sum_{\gamma \subset \partial K' \cap \partial K} \beta_\gamma \int_K \psi_\gamma \operatorname{sgn}(\overline{r})\, d\boldsymbol{x}. \tag{2.104}$$

Arguing similarly to above, it is readily concluded that α_K is uniformly bounded above and below by constants independent of $h_{K'}$.

Lemma 2.9 *Let $K' \in \mathcal{P}$ and define $v \in H_0^1(\Omega)$ as in (2.99). Then, for all elements $K \subset \widetilde{K}'$ we have*

$$h_K^{-2} \|v\|_{L_1(K)} + \|\Delta v\|_{L_1(K)} \leq C \tag{2.105}$$

and

$$h_K^{-1} \|v\|_{L_1(\partial K)} + \left\| \frac{\partial v}{\partial n} \right\|_{L_1(\partial K)} \leq C \tag{2.106}$$

where C is a positive constant independent of h_K.

Proof. First, observe that

$$\|v\|_{L_\infty(\widetilde{K}')} \leq C \left\{ \sum_{K \subset \widetilde{K}'} |\alpha_K| + \sum_{\gamma \subset \partial K'} |\beta_\gamma| \right\} \leq C$$

since $|\alpha_K|$ and $|\beta_\gamma|$ are uniformly bounded and where the constant C only depends on the regularity of the elements in the patch. Now,

$$\|v\|_{L_1(K)} \leq C h_K^2 \|v\|_{L_\infty(K)} \leq C h_K^2$$

and, furthermore, thanks to an inverse estimate (see Theorem 1.3), we obtain

$$\|\Delta v\|_{L_1(K)} \leq C h_K^{-2} \|v\|_{L_1(K)},$$

and the first estimate follows at once. The second estimate is obtained using similar arguments. ■

Observe that since $\overline{r}$ is piecewise constant, we obtain

$$\sum_{K \subset \widetilde{K}'} \int_K \overline{r} v \, \mathrm{d}\boldsymbol{x} = \sum_{K \subset \widetilde{K}'} \|\overline{r}\|_{L_\infty(K)} \int_K v \, \mathrm{sgn}(\overline{r}) \, \mathrm{d}\boldsymbol{x} \tag{2.107}$$

and, similarly,

$$\sum_{\gamma \subset \partial K'} \int_\gamma \overline{R} v \, \mathrm{d}s = \sum_{\gamma \subset \partial K'} \|\overline{R}\|_{L_\infty(\gamma)} \int_\gamma v \, \mathrm{sgn}(\overline{R}) \, \mathrm{d}s. \tag{2.108}$$

Then, thanks to the properties (2.100)–(2.101) characterizing the function v, these may be rewritten as

$$\sum_{K \subset \widetilde{K}'} h_K^2 \|\overline{r}\|_{L_\infty(K)} \tag{2.109}$$

and

$$\sum_{\gamma \subset \partial K'} h_\gamma \left\| \overline{R} \right\|_{L_\infty(\gamma)} \tag{2.110}$$

respectively. Consequently, it follows that

$$\eta(K')_\infty = \sum_{K \subset \widetilde{K}'} \int_K \overline{r} v \, \mathrm{d}\boldsymbol{x} + \sum_{\gamma \subset \partial K'} \int_\gamma \overline{R} v \, \mathrm{d}s \tag{2.111}$$

where

$$\eta(K')_\infty = \sum_{K \subset \widetilde{K}'} h_K^2 \left\| \overline{r} \right\|_{L_\infty(K)} + \sum_{\gamma \subset \partial K' \setminus \partial \Omega} h_\gamma \left\| \overline{R} \right\|_{L_\infty(\gamma)} . \tag{2.112}$$

Furthermore, $v \in H_0^1(\Omega)$; and so identity (2.9), along with the fact that v is supported on the patch $\widetilde{K}'$, leads to the conclusion

$$B(e, v) = \sum_{K \subset \widetilde{K}'} \int_K r v \, \mathrm{d}\boldsymbol{x} + \sum_{\gamma \subset \partial K'} \int_\gamma R v \, \mathrm{d}s \tag{2.113}$$

and therefore

$$\eta(K')_\infty = B(e, v) + \sum_{K \subset \widetilde{K}} \int_K (\overline{r} - r) v \, \mathrm{d}\boldsymbol{x} + \sum_{\gamma \subset \partial K} \int_\gamma (\overline{R} - R) v \, \mathrm{d}s. \tag{2.114}$$

Observing that for a first-order finite element approximation on triangular elements, $\overline{R} = R$ and $\overline{r} - r = \overline{f} - f$, we conclude that

$$\eta(K')_\infty = B(e, v) + \sum_{K \subset \widetilde{K}} \int_K (\overline{f} - f) v \, \mathrm{d}\boldsymbol{x}. \tag{2.115}$$

The first term on the right-hand side is rewritten using integration by parts as follows:

$$B(e, v) = \sum_{K \in \mathcal{P}} \int_K \nabla e \cdot \nabla v \, \mathrm{d}\boldsymbol{x} = \sum_{K \in \mathcal{P}} \left(- \int_K e \Delta v \, \mathrm{d}\boldsymbol{x} + \int_{\partial K} e \frac{\partial v}{\partial n} \, \mathrm{d}s \right),$$

and then, by Hölder's inequality, we obtain

$$|B(e, v)| \leq \sum_{K \in \widetilde{K}'} \left\{ \|e\|_{L_\infty(K)} \|\Delta v\|_{L_1(K)} + \|e\|_{L_\infty(\partial K)} \left\| \frac{\partial v}{\partial n} \right\|_{L_1(\partial K)} \right\}, \tag{2.116}$$

while the second term is bounded using the estimate

$$\left| \int_K (\overline{f} - f) v \, \mathrm{d}\boldsymbol{x} \right| \leq \left\| \overline{f} - f \right\|_{L_\infty(K)} \|v\|_{L_1(K)} . \tag{2.117}$$

Applying Lemma 2.9 leads to the conclusion

$$\eta(K')_\infty \leq C\,\|e\|_{L_\infty(\widetilde{K}')} + C \sum_{K \subset \widetilde{K}'} h_K^2 \,\|\overline{f} - f\|_{L_\infty(K)}. \tag{2.118}$$

The developments in this, and the previous section, are recorded as a Theorem.

Theorem 2.10 *Suppose that the assumptions of Lemma 2.8 hold and let η_∞ be defined as in (2.97). Then, the error e in the piecewise linear finite element approximation satisfies*

$$\|e\|_{L_\infty(\Omega)} \leq C|\log h_{max}|^2 \eta_\infty \tag{2.119}$$

and

$$\eta_\infty \leq C\,\|e\|_{L_\infty(\Omega)} + C \max_{K \in \mathcal{P}} h_K^2 \,\|\overline{f} - f\|_{L_\infty(K)} \tag{2.120}$$

where C is a positive constant independent of the mesh size and $\overline{f}$ is a piecewise constant approximation to f.

Proof. The first bound is a repeat of (2.98). Let K' be an element such that $\eta(K')_\infty$ is the largest. The second bound follows from the estimate (2.118) by arguing that

$$\eta_\infty \leq C\eta(K')_\infty \leq C\,\|e\|_{L_\infty(\widetilde{K}')} + C \sum_{K \subset \widetilde{K}'} h_K^2 \,\|\overline{f} - f\|_{L_\infty(K)}$$

and noting that

$$\|e\|_{L_\infty(\widetilde{K}')} \leq \|e\|_{L_\infty(\Omega)}$$

and

$$\sum_{K \subset \widetilde{K}'} h_K^2 \,\|\overline{f} - f\|_{L_\infty(K)} \leq C \max_{K \in \mathcal{P}} h_K^2 \,\|\overline{f} - f\|_{L_\infty(K)}$$

as required. ■

2.6 BIBLIOGRAPHICAL REMARKS

Explicit error estimators were derived originally by Babuška and Rheinboldt [24] for problems posed in one space dimension. The two-dimensional case was considered in the work of Babuška and Miller [19] and by Kelly *et al.* [74]. The technique of using bubble functions to obtain two-sided bounds on the error was developed by Verfürth [111]. The pointwise estimator was analyzed by Nochetto [85]. The extension of the pointwise estimator to three dimensions will be found in reference [61].

3

Implicit A Posteriori Estimators

3.1 INTRODUCTION

Error estimators that may be computed directly from the finite element approximation and the data for the problem of interest are referred to as *explicit estimators.* The estimators discussed in this chapter require the (approximate) solution of an auxiliary boundary value problem whose solution is notionally an approximation to the actual error *function*, as opposed to a norm of the error, and are collectively referred to as *implicit estimators.*

Why is it worthwhile considering implicit estimators? The explicit estimators presented in the previous chapter are typically comprised of residual terms of three distinct types. The interior residual reflects how well the finite element approximation satisfies the underlying partial differential equation on the interior of the domain, while the residual at a Neumann boundary measures the accuracy of the approximation of the boundary condition. The interelement residuals depend on the jumps in the numerical approximation of the normal fluxes at the element boundaries and reflect the regularity of the approximation. While each type of residual was scaled with an appropriate power of the local mesh size, the multiplicative constants were, in the absence of any simple alternative, absorbed into a single constant C, whose value is generally unknown. Moreover, the three types of residual stemmed from the use of the triangle inequality on the residual equation, followed by various applications of the Cauchy–Schwarz inequality. The danger with this explicit approach is that any possible cancellation between the various types of residual is lost by separating terms using the triangle inequality. Similarly, information is lost through applying the Cauchy–Schwarz inequality or simply lumping the three types of term together under a single (unknown) multiplicative factor. One may, of course, obtain suitable bounds for the value of the constant. However, the values that are dictated by the most pathological functions that could arise and so, for the most part, are somewhat pessimistic. Indeed, at each stage of the argument, the estimates are sharp only in the worst-case scenario, so that, overall the resulting estimators tend to be pessimistic and fail to detect the more subtle nuances of the specific problem at hand. Implicit estimators seek to avoid these disadvantages by retaining the structure of the original equation as far as possible. To wit, a local boundary value problem is solved with the residuals as data. This curtails the need for generic constants and the correct balance between the types of residual is accounted for by the solution process itself. The drawback is that one is obliged to solve an additional boundary value problem.

The true error satisfies the residual equation

$$B(e, v) = B(u, v) - B(u_X, v) = L(v) - B(u_X, v) \quad \forall v \in V. \tag{3.1}$$

In principle, one could approximate the problem (3.1) and obtain an approximation $\bar{e}$ to the actual error function. Unfortunately, reusing the original finite element subspace X to produce an approximation would result in a trivial solution, $\bar{e} = 0$. Consequently, a larger space than X is required, in which case the expense entailed would presumably be better directed toward producing a superior approximation of the original problem, rather than simply estimating error.

An alternative is to replace the global problem by sequence of uncoupled local boundary valuc problcms that approximate the single global residual equation, in an appropriate sense. The boundary value problems are local in that they are posed either over a single element (the *element residual method*) or over a small patch of elements (thc *subdomain rcsidual method*). The error estimator is obtained by evaluating the norms of the solutions of the local problems and summing contributions over the domain. The ideas are illustrated by considering approximation of the model problem in Section 1.4 using a finite element space constructed on a regular partition of the domain Ω.

3.2 THE SUBDOMAIN RESIDUAL METHOD

The first method that we will consider consists of decomposing the global residual problem for the error into a sequence of local residual problems with *homogeneous essential boundary data* posed over small patches or subdomains of the domain Ω.

3.2.1 Formulation of Subdomain Residual Problem

Let $\mathcal{N}$ index the set of element vertices in the partition $\mathcal{P}$ and let $\{\theta_n\}_{n\in\mathcal{N}}$ denote the first-order Lagrange basis functions based at the element vertices. These functions are characterized by the conditions

$$\theta_n(\boldsymbol{x}_m) = \delta_{nm} \tag{3.2}$$

where $\boldsymbol{x}_m$ is any vertex in $\mathcal{N}$ and

$$\sum_{n\in\mathcal{N}} \theta_n(\boldsymbol{x}) \equiv 1, \quad \boldsymbol{x} \in \overline{\Omega}. \tag{3.3}$$

The support $\widetilde{\Omega}_n$ of the nodal function θ_n consists of the patch of elements containing the vertex $\boldsymbol{x}_n$.

The method is formulated starting with the equation (3.1) characterizing the true error e. While, at least in principle, one could approximate the solution of this equation by solving the problem using an enhanced finite element subspace X, it would be simpler to compute a new finite element approximation directly. The underlying idea is to replace the single global problem characterizing the error, by a sequence of independent problems posed on small subdomains of the partition $\mathcal{P}$. The nodal basis functions may be utilized for this purpose. With the aid of property (3.3), we find that

$$B(e,v) = B(e, v\sum_{n\in\mathcal{N}}\theta_n) = \sum_{n\in\mathcal{N}} B(e, v\theta_n) = \sum_{n\in\mathcal{N}} L(v\theta_n) - B(u_X, v\theta_n) \tag{3.4}$$

for all $v \in V$. The function $\theta_n v$ is supported on the set $\widetilde{\Omega}_n$ and vanishes on the boundary. Therefore, it follows that $\theta_n v$ belongs to the space $H_0^1(\widetilde{\Omega}_n)$. The local bilinear form associated with this space $B_n : H_0^1(\widetilde{\Omega}_n) \times H_0^1(\widetilde{\Omega}_n) \mapsto \mathbb{R}$ is given by

$$B_n(u,v) = \int_{\widetilde{\Omega}_n} (\boldsymbol{\nabla} u \cdot \boldsymbol{\nabla} v + cuv)\, \mathrm{d}\boldsymbol{x} \tag{3.5}$$

and the local load functional $L_n : H_0^1(\widetilde{\Omega}_n) \mapsto \mathbb{R}$ is given by

$$L_{\widetilde{\Omega}_n}(v) = \int_{\widetilde{\Omega}_n} fv\, \mathrm{d}\boldsymbol{x} + \int_{\partial\widetilde{\Omega}_n \cap \Gamma_N} gv\, \mathrm{d}s. \tag{3.6}$$

The *subdomain residual problem* consists of finding $\phi_n \in H_0^1(\widetilde{\Omega}_n)$ such that

$$B_n(\phi_n, v) = L_n(v) - B_n(u_X, v) \quad \forall v \in H_0^1(\widetilde{\Omega}_n). \tag{3.7}$$

Motivated by equation (3.4), the error estimator η_n associated with the subdomain $\widetilde{\Omega}_n$ is taken to be

$$\eta_n = |\!|\!| \phi_n |\!|\!|_{\widetilde{\Omega}_n} \tag{3.8}$$

and the global error estimator η is obtained by summing the contributions from the subdomains

$$\eta = \left\{ \sum_{n \in \mathcal{N}} \eta_n^2 \right\}^{1/2} . \tag{3.9}$$

In practice, the subdomain residual problems are approximated using a finite-dimensional subspace of $H_0^1(\widetilde{\Omega}_n)$. This aspect will be discussed further in Section 3.2.4.

3.2.2 Preliminaries

Suppose that the basis function θ_n is nonzero on an element K; then, applying Lemma 1.4, there exists a constant C depending only on the regularity of the element such that

$$|\theta_n(\boldsymbol{x})| \leq C, \quad \boldsymbol{x} \in K \tag{3.10}$$

and

$$|\boldsymbol{\nabla}\theta_n(\boldsymbol{x})| \leq C h_K^{-1}, \quad \boldsymbol{x} \in K. \tag{3.11}$$

It is possible to partition the set of vertices $\mathcal{N}$ into the union of disjoint subsets $\mathcal{N}_1, \mathcal{N}_2, \ldots$ such that any pair of nodal basis functions in the same subset $\mathcal{N}_r$ have nonoverlapping supports. Specifically, the condition that will be required is

$$\forall m, n \in \mathcal{N}_r : m \neq n \Rightarrow \quad \text{int supp}\, \theta_n \cap \text{int supp}\, \theta_m \text{ is empty} \tag{3.12}$$

where int supp θ_n denotes the interior of the support of θ_n. It is easy to see that such a partitioning exists since one can simply choose each of the sets $\mathcal{N}_r$ to consist of a single vertex. However, it will later be seen that using as few subsets as possible is most advantageous. The smallest possible number of subsets is denoted by ν and referred to as the *overlap index* for the partition. The overlap index for a regular family of partitions may be bounded by a constant depending only on the regularity κ of the elements in the family.

Let K be any element in the partition $\mathcal{P}$. The set of element basis functions that are nonzero on this element is denoted by $\mathcal{N}(K)$. The maximum cardinality of these sets is denoted by τ and referred to as the *intersection index*. If the partition is regular, then the intersection index will coincide with the maximum number of vertices in any element and is therefore uniformly bounded for a regular family of partitions.

As before, recall (1.30), the subdomain consisting of the elements neighboring element K, is denoted by

$$\widetilde{K} = \left\{ K' \in \mathcal{P} : \overline{K}' \cap \overline{K} \text{ is nonempty} \right\}. \tag{3.13}$$

Thanks to the regularity assumption on the elements, it follows that the maximum number of elements contained in any of these subdomains is bounded by a multiple of τ.

The approximation properties of the finite element subspaces will be required in the analysis. In particular, applying Theorem 1.7, it follows at once that there exists a constant C such that for any $v \in H^1(\Omega)$ and any element K there holds

$$h_K^{-1} \|v - \mathcal{I}_X v\|_{L_2(K)} + |v - \mathcal{I}_X v|_{H^1(K)} \leq C \|v\|_{H^1(\tilde{K})} \tag{3.14}$$

where $\mathcal{I}_X$ is the regularized finite element approximation operator in Theorem 1.7.

The above arguments are valid regardless of the degree p of the polynomial subspace actually used in the construction of the finite element subspace. This follows since it is only the first-order Lagrange basis functions that appear in the discussion. As a consequence we merely need to impose restrictions on the mesh topology and mesh geometry. While it is sufficient to assume that the family of partitions is regular, this is by no means necessary and it is possible to extend the analysis to more general classes of partition [19].

3.2.3 Equivalence of Estimator

Lemma 3.1 *There exists a constant C depending only on the regularity of the elements such that for any $v \in H^1(\Omega)$ we have*

$$\sum_{n \in \mathcal{N}} \|\theta_n(v - \mathcal{I}_X v)\|^2_{H^1(\Omega)} \leq C\tau^2 \|v\|^2_{H^1(\Omega)} \tag{3.15}$$

where τ is the intersection index.

Proof. Let K be any element and let $\theta_n \in \mathcal{N}$ be a nodal basis function that does not vanish on K. Then,

$$\begin{aligned} &\|\theta_n(v - \mathcal{I}_X v)\|^2_{H^1(K)} \\ &\leq \|\theta_n\|^2_{L_\infty(K)} |v - \mathcal{I}_X v|^2_{H^1(K)} + |\theta_n|^2_{W^{1,\infty}(K)} \|v - \mathcal{I}_X v\|^2_{L_2(K)} . \end{aligned}$$

Thanks to properties (3.10)–(3.11), it follows that

$$\|\theta_n(v - \mathcal{I}_X v)\|^2_{H^1(K)} \leq C \left\{ |v - \mathcal{I}_X v|^2_{H^1(K)} + h_K^{-2} \|v - \mathcal{I}_X v\|^2_{L_2(K)} \right\};$$

hence, with the aid of the approximation property (3.14), we conclude

$$\|\theta_n(v - \mathcal{I}_X v)\|^2_{H^1(K)} \leq C \|v\|^2_{H^1(\tilde{K})} .$$

Summing this inequality over all vertices $n \in \mathcal{N}$ gives the result

$$\sum_{n \in \mathcal{N}} \|\theta_n(v - \mathcal{I}_X v)\|^2_{H^1(\Omega)} = \sum_{K \in \mathcal{P}} \sum_{n \in \mathcal{N}(K)} \|\theta_n(v - \mathcal{I}_X v)\|^2_{H^1(K)}$$

$$
\begin{aligned}
&\le C \sum_{K\in\mathcal{P}} \sum_{n\in\mathcal{N}(K)} \|v\|^2_{H^1(\widetilde{K})} \\
&= C\tau \sum_{K\in\mathcal{P}} \|v\|^2_{H^1(\widetilde{K})} \\
&\le C\tau^2 \|v\|^2_{H^1(\Omega)}
\end{aligned}
$$

as claimed. ■

Theorem 3.2 *Let η denote the error estimator (3.9) obtained using the subdomain residual method. Then there exists a constant C depending only on the regularity of the elements such that*

$$\frac{\eta}{\sqrt{\nu}} \le |\!|\!| e |\!|\!| \le C\tau\eta, \tag{3.16}$$

where τ is the overlap index and ν is the intersection index.

Proof. Using the Galerkin orthogonality, we obtain

$$B(e,v) = B(e, v - \mathcal{I}_X v)$$

and then by property (3.3) we have

$$B(e,v) = \sum_{n\in\mathcal{N}} B(e, \theta_n(v - \mathcal{I}_X v)).$$

Noting that $\theta_n(v - \mathcal{I}_X v) \in H^1_0(\widetilde{\Omega}_n)$, we obtain

$$B(e,v) = \sum_{n\in\mathcal{N}} B(\phi_n, \theta_n(v - \mathcal{I}_X v))$$

where ϕ_n is defined in (3.7). Hence

$$|B(e,v)| \le \sum_{n\in\mathcal{N}} \eta_n \, |\!|\!| \theta_n(v - \mathcal{I}_X v) |\!|\!| \, .$$

Applying the Cauchy–Schwarz inequality and Lemma 3.1, we have

$$|B(e,v)| \le C\eta \left\{ \sum_{n\in\mathcal{N}} \|\theta_n(v - \mathcal{I}_X v)\|^2_{H^1(\Omega)} \right\}^{1/2} \le C\eta\tau \|v\|_{H^1(\Omega)} \le C\eta\tau \, |\!|\!| v |\!|\!|$$

from where there follows

$$|\!|\!| e |\!|\!| \le C\tau\eta.$$

Conversely, consider

$$\begin{aligned}\eta^2 &= \sum_{n\in\mathcal{N}} \eta_n^2 = \sum_{n\in\mathcal{N}} B_n(\phi_n,\phi_n) \\ &= \sum_{n\in\mathcal{N}} B_n(e,\phi_n) = \sum_{n\in\mathcal{N}} B(e,\phi_n) \\ &= B(e, \sum_{n\in\mathcal{N}} \phi_n) \\ &\leq |||e||| \left|\left|\left|\sum_{n\in\mathcal{N}} \phi_n\right|\right|\right| .\end{aligned}$$

By partitioning the set $\mathcal{N}$ of vertices as described above, we obtain

$$\left|\left|\left|\sum_{n\in\mathcal{N}} \phi_n\right|\right|\right|^2 = \left|\left|\left|\sum_{\mathcal{N}_r}\sum_{n\in\mathcal{N}_r} \phi_n\right|\right|\right|^2 \leq \nu \sum_{\mathcal{N}_r} \left|\left|\left|\sum_{n\in\mathcal{N}_r} \phi_n\right|\right|\right|^2 = \nu \sum_{\mathcal{N}_r}\sum_{n\in\mathcal{N}_r} |||\phi_n|||^2$$

where property (3.12) has been used. Consequently,

$$\left|\left|\left|\sum_{n\in\mathcal{N}} \phi_n\right|\right|\right|^2 \leq \nu\eta^2$$

and the result follows immediately. ■

3.2.4 Treatment of Residual Problems

The subdomain residual method described above requires the *true* solution of the local problems

$$\phi_n \in H_0^1(\widetilde{\Omega}_n) : B_n(\phi_n, v_n) = L(v_n) - B_n(u_X, v_n) \quad \forall v_n \in H_0^1(\widetilde{\Omega}_n). \tag{3.17}$$

Of course, it is infeasible to require the exact solution of the problem. However, even if one were satisfied with an approximation, in practice, the method is seldom used, partly because it is inconvenient and relatively expensive to develop approximations over the irregularly shaped patches $\widetilde{\Omega}_n$.

Babuška and Miller [19] circumvented this difficulty by obtaining an equivalent measure for the local error estimator $|||\phi_n|||$ as follows. The basic idea is reminiscent of the explicit estimators of Chapter 2. Integrating the right-hand side of (3.17) by parts gives, for all $v_n \in H_0^1(\widetilde{\Omega}_n)$,

$$L(v_n) - B_n(u_X, v_n) = \sum_{K\in\widetilde{\Omega}_n} \left\{ \int_K r v_n \, \mathrm{d}\boldsymbol{x} + \sum_{\gamma\subset\widetilde{\Omega}_n} \int_\gamma R v_n \, \mathrm{d}s \right\} \tag{3.18}$$

where r and R are the usual interior and boundary residuals. A routine application of the Cauchy–Schwarz inequality leads to the bound

$$\eta_n \leq C \left\{ \sum_{K \subset \tilde{\Omega}_n} h_K^2 \, \|r\|_{L^2(K)}^2 + \sum_{\gamma \subset \tilde{\Omega}_n} h_K \, \|R\|_{L^2(\gamma)}^2 \right\}^{1/2} . \tag{3.19}$$

Summing this estimate over all vertices in the partition and regrouping the terms leads to the familiar *explicit error estimator*

$$\|\!|e|\!\| \leq C \sum_{K \in \mathcal{P}} \eta_K^2, \tag{3.20}$$

where

$$\eta_K^2 = \|r\|_{L^2(K)}^2 + \sum_{\gamma \subset \partial K} h_K \, \|R\|_{L^2(\gamma)}^2 . \tag{3.21}$$

Babuška and Miller also obtain the following result:

$$\sum_{K \subset \tilde{\Omega}_n} h_K \, \|r\|_{L^2(K)} + \sum_{\gamma \subset \tilde{\Omega}_n} h_K^{1/2} \, \|R\|_{L^2(\gamma)} \leq C(\eta_n + \epsilon) \tag{3.22}$$

where

$$\epsilon = \sum_{K \subset \tilde{\Omega}_n} h_K \, \|r - \overline{r}\|_{L^2(K)} + \sum_{\gamma \subset \tilde{\Omega}_n} h_K^{1/2} \, \|R - \overline{R}_\gamma\|_{L^2(\gamma)} \tag{3.23}$$

and $\overline{r}$ and $\overline{R}$ are finite-dimensional approximations to r and R as described in Section 2.3. This result is essentially identical to those derived in Section 2.3 when considering explicit error estimators, but predates the discovery of those estimates.

3.3 THE ELEMENT RESIDUAL METHOD

3.3.1 Formulation of Local Residual Problem

One of the main disadvantages of the subdomain residual method is that the local patch problems are rather expensive to approximate accurately. In effect, each element is treated several times according to the number of patches with which it is associated. A natural alternative is to attempt to reduce the cost by solving local problems posed over individual elements.

Let the error on an element K be denoted by $e = u - u_X$. Thanks to the smoothness of the finite element approximation on the element interiors, one finds that the error satisfies the differential equation

$$-\Delta e + ce = f + \Delta u_X - cu_X \text{ in } K. \tag{3.24}$$

The major difficulty is to supplement the equation with appropriate boundary conditions. There are various cases to consider. First, suppose that the element K intersects a portion of the boundary of the domain Ω where an essential boundary condition is imposed. The appropriate boundary condition for the local error residual problem is clearly

$$e = 0 \text{ on } \partial K \cap \Gamma_D. \tag{3.25}$$

Here, it has been assumed that the finite element approximation has been constructed so that the Dirichlet boundary conditions are satisfied exactly (although this is not essential).

Next, suppose that the element intersects a portion of the boundary $\partial\Omega$ where a natural boundary condition is imposed. The local error residual problem is subjected to a natural boundary condition

$$\frac{\partial e}{\partial n_K} = g - \frac{\partial u_X}{\partial n_K} \text{ on } \partial K \cap \Gamma_N. \tag{3.26}$$

So far, appropriate boundary conditions have been clear. It remains to consider the case when the element boundary lies on the interior of the domain.

Dealing with individual elements renders the use of homogeneous boundary conditions for the local problems much less attractive, since the entire set of local solutions would vanish on the interelement edges. Instead, the *element residual method* imposes nonhomogenous flux boundary conditions at the element boundaries. Ideally, one would like to use the exact data and impose the condition

$$\frac{\partial e}{\partial n_K} = \frac{\partial u}{\partial n_K} - \left.\frac{\partial u_X}{\partial n_K}\right|_K \tag{3.27}$$

on the edge separating elements K and J, where $u_X|_K$ denotes the restriction of the finite element approximation to an element K. Of course, the true flux appearing on the right-hand side is in general unknown. However, one may replace the true flux by an approximation obtained from the finite element solution itself:

$$\frac{\partial u}{\partial n_K} \approx \left\langle \frac{\partial u_X}{\partial n_K} \right\rangle \tag{3.28}$$

where $\langle\cdot\rangle$ denotes the averaged flux:

$$\left\langle \frac{\partial u_X}{\partial n_K} \right\rangle = \frac{1}{2}\boldsymbol{n}_K \cdot \{(\boldsymbol{\nabla} u_X)_K + (\boldsymbol{\nabla} u_X)_J\}. \tag{3.29}$$

The idea behind this choice is that by averaging the discontinuous finite element approximation to the normal flux, one might obtain a sufficiently accurate approximation to the true flux.

The error residual problem is formulated as a variational equation as follows. On each element K, the actual error satisfies the boundary value problem

$$B_K(e,v) = F_K(v) - B_K(u_X,v) + \int_{\partial K} \frac{\partial u}{\partial n_K} v \, ds \quad \forall v \in V_K. \tag{3.30}$$

Here,

$$V_K = \{v \in H^1(K) : v = 0 \text{ on } \partial K \cap \Gamma_D\} \tag{3.31}$$

and $B_K : V_K \times V_K \mapsto \mathbb{R}$ is the local bilinear form

$$B_K(u, v) = \int_K (\nabla u \cdot \nabla v + cuv) \, \mathrm{d}\boldsymbol{x} \tag{3.32}$$

and $F_K : V_K \mapsto \mathbb{R}$ is the local load functional

$$F_K(v) = \int_K fv \, \mathrm{d}\boldsymbol{x}. \tag{3.33}$$

The *error residual problem* is the weak form of the problem specified by the conditions (3.24)–(3.28): Find $\phi_K \in V_K$ such that

$$B_K(\phi_K, v) = F_K(v) - B_K(u_X, v) + \int_{\partial K} \left\langle \frac{\partial u_X}{\partial n_K} \right\rangle v \, \mathrm{d}s \quad \forall v \in V_K. \tag{3.34}$$

Here, the definition of the average has been extended to include the exterior boundary:

$$\left\langle \frac{\partial u_X}{\partial n_K} \right\rangle = \begin{cases} \frac{1}{2}\boldsymbol{n}_K \cdot \{(\nabla u_X)_K + (\nabla u_X)_{K'}\} & \text{on } \partial K \cap \partial K' \\ \boldsymbol{n}_K \cdot (\nabla u_X)_K & \text{on } \partial K \cap \Gamma_D \\ g & \text{on } \partial K \cap \Gamma_N. \end{cases} \tag{3.35}$$

3.3.2 Solvability of the Local Problems

The local problem (3.34) is subjected to pure Neumann boundary conditions. Does a solution a exist? In general, the answer is no! The "difficulty" arises from the nontrivial kernel associated with the bilinear form $B_K(\cdot, \cdot)$. Unless the data satisfies appropriate compatibility conditions, the problem will fail to possess solutions. Ideally, one should try to choose the boundary data more carefully so that the underlying problem is naturally well-posed. However, the coupling between neighboring elements means that this is easier said than done. Nevertheless, this approach (the *equilibrated residual method*) is pursued in Chapter 6.

An alternative possibility is to reformulate the approximate problem over a subspace Y_K of V_K on which the bilinear form will be coercive. This ostensibly crude approach, if *carefully* applied, leads to useful error estimators. However, constructing subspaces Y_K through tinkering with the full space V_K is a risky business, particularly in the case of higher-order finite element approximation. Guidelines for choosing the subspace Y_K are well established in the case of first-order finite element approximation. Consequently, the discussion for the remainder of this section is restricted to first-order finite element approximation. Treatment of the general case will be resumed in Section 3.4.2.

Bilinear Approximation on Quadrilaterals Consider the case of finite element approximation using piecewise bilinear functions on quadrilateral elements. The local approximation space for the error residual problem is constructed using the nodal basis functions specified on the reference element

$$\widehat{K} = \{(\widehat{x}, \widehat{y}) : -1 \leq \widehat{x} \leq 1; \quad -1 \leq \widehat{y} \leq 1\}. \tag{3.36}$$

The edge bubble functions are given by

$$\left.\begin{aligned} \widehat{\chi}_1 &= \tfrac{1}{2}(1 - \widehat{x}^2)(1 - \widehat{y}) \\ \widehat{\chi}_2 &= \tfrac{1}{2}(1 - \widehat{y}^2)(1 + \widehat{x}) \\ \widehat{\chi}_3 &= \tfrac{1}{2}(1 - \widehat{x}^2)(1 + \widehat{y}) \\ \widehat{\chi}_4 &= \tfrac{1}{2}(1 - \widehat{y}^2)(1 - \widehat{x}) \end{aligned}\right\} \tag{3.37}$$

and the interior bubble function $\widehat{\psi}$ is given by

$$\widehat{\psi} = (1 - \widehat{x}^2)(1 - \widehat{y}^2). \tag{3.38}$$

The bubble space $\widehat{Y}$ is defined by

$$\widehat{Y} = \text{span}\left\{\widehat{\chi}_1, \widehat{\chi}_2, \widehat{\chi}_3, \widehat{\chi}_4, \widehat{\psi}\right\} \tag{3.39}$$

and the error residual problem is approximated using the subspace Y_K obtained by mapping the space $\widehat{Y}$ to the element K.

Piecewise Affine Approximation on Triangles Consider the case of finite element approximation using piecewise affine functions on linear triangular elements. The local approximation space for the error residual problem is constructed using the basis functions specified on the reference element

$$\widehat{K} = \{(\widehat{x}, \widehat{y}) : 0 \leq \widehat{x} \leq 1; \quad 0 \leq \widehat{y} \leq 1 - \widehat{x}\} \tag{3.40}$$

The functions $\widehat{\lambda}_1$, $\widehat{\lambda}_2$, and $\widehat{\lambda}_3$ denote the barycentric (area) coordinates on $\widehat{K}$ and the edge bubble functions are given by

$$\widehat{\chi}_1 = 4\widehat{\lambda}_2\widehat{\lambda}_3; \quad \widehat{\chi}_2 = 4\widehat{\lambda}_1\widehat{\lambda}_3; \quad \widehat{\chi}_3 = 4\widehat{\lambda}_1\widehat{\lambda}_2. \tag{3.41}$$

Let $\widehat{Y}$ denote the bubble space

$$\widehat{Y} = \text{span}\left\{\widehat{\chi}_1, \widehat{\chi}_2, \widehat{\chi}_3\right\}. \tag{3.42}$$

The error residual problem is approximated using the subspace Y_K obtained by mapping the space $\widehat{Y}$ to the element K. This choice of space contains no interior bubble function. An alternative choice [109] is to include the extra interior bubble function

$$\widehat{\chi}_4 = \widehat{\lambda}_1\widehat{\lambda}_2\widehat{\lambda}_3. \tag{3.43}$$

3.3.3 The Classical Element Residual Method

Once a suitable finite-dimensional subspace Y_K has been selected, the classical element residual method [34, 37, 55, 89] for *a posteriori* error estimation consists of computing the solution of the local error residual problems

$$\phi_K \in Y_K: \quad B_K(\phi_K, v) = F_K(v) - B_K(u_X, v) + \int_{\partial K} \left\langle \frac{\partial u_X}{\partial n_K} \right\rangle v \, \mathrm{d}s \quad \forall v \in Y_K, \tag{3.44}$$

thereby obtaining a function ϕ_K. The local error estimator η_K on element K is then defined by

$$\eta_K = |||\phi_K|||_K \tag{3.45}$$

and the global error estimator is obtained by summing the local contributions

$$|||e||| \approx \eta = \left\{ \sum_{K \in \mathcal{P}} \eta_K^2 \right\}^{1/2}. \tag{3.46}$$

3.3.4 Relationship with Explicit Error Estimators

The implicit estimators can be related to the explicit estimators of Chapter 2. By applying Green's identity to the right-hand side of the error residual equation (3.44), one obtains

$$B_K(\phi_K, v) = \int_K r v \, \mathrm{d}\boldsymbol{x} + \int_{\partial K} R v \, \mathrm{d}s \quad \forall v \in Y_K \tag{3.47}$$

where r and R are the precisely the interior and boundary residuals appearing in the explicit error estimators. Choosing the function v to be the solution ϕ_K of the local problem leads to

$$\eta_K^2 = \int_K r \phi_K \, \mathrm{d}\boldsymbol{x} + \int_{\partial K} R \phi_K \, \mathrm{d}s. \tag{3.48}$$

By Poincaré's inequality, there exists a constant C which depends only on the shape regularity of the element K but not on its size such that

$$||\phi_K||_{L_2(K)} \leq C h_K \, |\phi_K|_{H^1(K)}, \tag{3.49}$$

and so for any edge $\gamma \subset \partial K$ we have

$$||\phi_K||_{L_2(\gamma)} \leq C h_K^{1/2} \, |\phi_K|_{H^1(K)}. \tag{3.50}$$

With the aid of these results and the Cauchy–Schwarz inequality, we deduce

$$\eta_K^2 \leq C \left\{ h_K \, ||r||_{L_2(K)} + \sum_{\gamma \subset \partial K} h_K^{1/2} \, ||R||_{L_2(\gamma)} \right\} |\phi_K|_{H^1(K)} \tag{3.51}$$

from where it follows that

$$\eta_K \leq C \left\{ h_K^2 \, \|r\|_{L_2(K)}^2 + \frac{1}{2} \sum_{\gamma \subset \partial K} h_K \, \|R\|_{L_2(\gamma)}^2 \right\}^{1/2} \tag{3.52}$$

where the term on the right-hand side is the explicit error estimator discussed in Chapter 2.

3.3.5 Efficiency and Reliability of the Estimator

The estimator produced by the element residual method provides two-sided bounds on the actual error similar to those obtained for explicit schemes.

For ease of exposition, suppose that the space Y_K used to approximate the error residual problem is equipped with an interior bubble function. Let $\psi_K \in Y_K$ denote the interior bubble constructed in Theorem 2.2. The following result complements Theorem 2.5:

Lemma 3.3 *Let r and R denote the interior and boundary residuals associated with the finite element approximation constructed from the subspace X. Suppose that $\overline{r}_K$ and $\overline{R}_\gamma$ are finite-dimensional approximations to the residuals on an element K and on an edge $\gamma \subset \partial K$ as in Chapter 2, and let η_K denote the element residual error estimator. Then there exists a constant C depending only on the shape regularity of the elements such that*

$$\|r\|_{L_2(K)} \leq C \left\{ h_K^{-1} \eta_K + \|\overline{r}_K - r\|_{L_2(K)} \right\} \tag{3.53}$$

and

$$\|R\|_{L_2(\gamma)} \leq C \left\{ h_K^{-1/2} \eta_K + h_K^{1/2} \, \|\overline{r}_K - r\|_{L_2(K)} + \left\|\overline{R}_\gamma - R\right\|_{L_2(\gamma)} \right\}. \tag{3.54}$$

Proof. The first result is shown by following the arguments from (2.40) to (2.49), except that the identity (obtained by choosing $v = \overline{r}_K \psi_K$ in (3.44))

$$B_K(\phi_K, \overline{r}_K \psi_K) = \int_K \psi_K r \overline{r}_K \, \mathrm{d}x$$

is used in place of (2.41). The second result is obtained by following arguments similar to those from (2.50) to (2.56) and, instead of (2.51), using the identity

$$B_K(\phi_K, \overline{R}_\gamma \chi_\gamma) = \int_K r \overline{R}_\gamma \chi_\gamma \, \mathrm{d}x + \int_\gamma \chi_\gamma R \overline{R}_\gamma \, \mathrm{d}s$$

which follows from (3.44) with the choice $v = \overline{R}_\gamma \chi_\gamma$. ■

Thanks to the relationship between the explicit and implicit error estimators, one immediately obtains the following result as a corollary of Theorem 2.6:

Theorem 3.4 *Let η_K denote the local error estimator obtained using the element residual method. Then there exists a constant C depending only on the shape regularity of the elements such that*

$$\|\!|e|\!\|^2 \le C \sum_{K\in\mathcal{P}} \left\{ \eta_K^2 + h_K^2 \left\|f - \overline{f}\right\|_{L_2(K)}^2 + \sum_{\gamma\subset\partial K\cap\Gamma_N} h_K \left\|g - \overline{g}\right\|_{L_2(\gamma)}^2 \right\}. \tag{3.55}$$

Moreover, the local lower bound

$$\eta_K^2 \le C \left\{ \|\!|e|\!\|_{\widetilde{K}}^2 + h_K^2 \left\|f - \overline{f}\right\|_{L_2(\widetilde{K})}^2 + \sum_{\gamma\subset\partial K\cap\Gamma_N} h_K \left\|g - \overline{g}\right\|_{L_2(\gamma)}^2 \right\} \tag{3.56}$$

holds for all elements $K \in \mathcal{P}$.

As noted elsewhere, the extra terms depend on the smoothness of the data and will generally be negligible in comparison with the estimator and the actual error. In this sense, the element residual method gives an equivalent measure of the discretization error in the energy norm.

3.4 THE INFLUENCE AND SELECTION OF SUBSPACES

It was observed in Section 3.3.2 that the local boundary value problem (3.34) leading to the element residual problem generally has no solution owing to the incompatibility of the data for the pure Neumann problem. However, reformulating the problem over a suitable subspace $Y_K \subset V_K$ may retrieve the situation. For instance, in the case of first-order finite element approximation on quadrilateral and triangular elements, the subspaces presented in Section 3.3.2 were shown to result in an *a posteriori* estimator with quite satisfactory properties. In both the case of bilinear approximation on quadrilateral elements and affine approximation on triangular elements, the same basic recipe applies for approximating the local problem (3.34): Increase the order of the space used to construct the original finite element approximation and then form the quotient space by subtracting the nodal finite element interpolant. While the selection of subspaces for first-order finite elements is reasonably well established, the situation is somewhat less clear for higher-order finite element approximation, and some rather nasty surprises are lurking.

3.4.1 Exact Solution of Element Residual Problem

To begin with, we consider the situation when the local residual problem is solved "exactly." Of course, this is hardly a practical proposition, but our purpose here is to isolate those effects not associated with approximation of the local problem. Here, the term "exact" is understood to mean that the space $Y_K = H^1(K)/\mathbb{R}$ is used in place of V_K to construct the solution of the local residual problem. The bilinear form $B_K(\cdot,\cdot)$ is coercive over the reduced space Y_K thanks to Poincaré's inequality. Consequently, the local residual problem (3.44) is well-posed over the space Y_K, and a solution $\phi_K \in Y_K$ exists and is unique.

Consider the classical element residual technique applied to pth-order finite element approximation on quadrilateral elements. The optimal rate of convergence in the energy norm that may be achieved using degree p elements is $O(h^p)$. Only in trivial cases can this rate be exceeded. If there exists a constant $C(u)$ for which the error in the energy is bounded below by $C(u)h^p$, then the error is said to be *properly* $O(h^p)$. If an estimator η converges to the true error in the limit $h \to 0$, then the estimator is said to be *asymptotically exact.* Clearly, this is a highly desirable property, but is usually not satisfied in practical situations. Nonetheless, under certain rather stringent circumstances, the element residual method is asymptotically exact when the local problems are solved exactly.

Asymptotic exactness on square elements of odd degree

Theorem 3.5 *Let $p \in \mathbb{N}$ be odd and assume $u \in H^{p+2}(\Omega)$. Suppose the error e measured in the energy norm is properly $O(h^p)$. Let η denote the error estimator obtained from the element residual method using the "exact" subspace $Y_K = H^1(K)/\mathbb{R}$. If the partition consists of square elements, then the estimator η is asymptotically exact,*

$$\lim_{h \mapsto 0} \frac{\eta}{|||e|||} = 1 \tag{3.57}$$

Proof. See reference [2]. ∎

The hypotheses in this result are sufficiently restrictive to rule out most situations of practical interest. However, the point is that such favorable circumstances are *sufficient* to ensure the element residual method used in conjunction with "exact" solution of the local problem produces an asymptotically exact estimator. What is perhaps more surprising is that these hypotheses are *necessary.*

Nonasymptotic exactness for elements of even order The estimator η is not asymptotically exact when the approximation is of even order as the following counterexample shows. Consider the following problem:

$$\begin{aligned} -\Delta u &= -p(p+1)x^{p-1} \text{ in } \Omega \\ u &= 0 \text{ on } \Gamma_D \\ \partial u/\partial n &= 0 \text{ on } \Gamma_N \end{aligned} \tag{3.58}$$

with $\Omega = (0,1) \times (0,1)$ and Γ_D being the vertical boundaries of Ω. The partition is formed by subdividing Ω into uniform squares of size h. The true solution is

$$u(x,y) = x(x^p - 1) \tag{3.59}$$

and the true error on element K may be computed explicitly [2]

$$\|e\|_K^2 = \frac{2h}{2p+1}\left(\frac{h}{2}\right)^{2p+1}\left(\frac{p+1}{k_p}\right)^2 \tag{3.60}$$

where k_p is the coefficient of the leading term in the Legendre polynomial. For the local error estimator η_K, it suffices to consider only those elements K lying on the interior of the partition, since the combined effect of elements on the boundary of the domain Ω becomes negligible as the partition is refined (provided the true solution is smooth). The estimator when the polynomial degree is even is

$$\eta_K^2 = h\left(\frac{h}{2}\right)^{2p+1}\left(\frac{p+1}{k_p}\right)^2\left\{\frac{2}{2p+1}+2\right\}. \tag{3.61}$$

Consequently, when p is even we have

$$\lim_{h\mapsto 0}\frac{\eta}{\|e\|} = \sqrt{2(p+1)}, \tag{3.62}$$

and therefore the estimator is no longer asymptotically exact.

Nonasymptotic exactness on nonsquare elements The estimator is not asymptotically exact when the subdomains are not square even if the approximation is of odd degree. Suppose that the domain Ω is a rhombus with acute angle θ. Details will be given for the case of first-order approximation of the problem:

$$-\Delta u = f \text{ in } \Omega \quad u = 0 \text{ on } \Gamma_D \quad \partial u/\partial n = g \text{ on } \Gamma_N. \tag{3.63}$$

The data f and g are chosen so that the true solution is $u(x,y) = y^2$. The domain Ω is partitioned into a mesh of $N \times N$ uniform rhombuses. The finite element approximation of this problem coincides with the nodal interpolant of the true solution. The true error is given by

$$\|e\|_K^2 = \frac{8h}{3}\left(\frac{h}{2}\sin\theta\right)^3. \tag{3.64}$$

The solution of the error residual problem is difficult to compute exactly for this example. However, bounds can be established using variational analysis [2]:

$$\|\phi\|_K^2 \geq h\left(\frac{h}{2}\sin\theta\right)^3\left\{\frac{8}{3}+\frac{16}{15}\cos^2\theta\right\}. \tag{3.65}$$

Hence, using the expression for the true error reveals

$$\frac{\|\phi\|_K^2}{\|e\|_K^2} \geq 1 + \frac{2}{5}\cos^2\theta. \tag{3.66}$$

Letting $\boldsymbol{n}_1$ and $\boldsymbol{n}_2$ denote unit outward normals on adjacent edges of an element, we obtain

$$\lim_{h \mapsto 0} \frac{\eta^2}{\|e\|^2} \geq 1 + \frac{2}{5} |\boldsymbol{n}_1 \cdot \boldsymbol{n}_2|^2 . \tag{3.67}$$

Therefore the estimator cannot be asymptotically exact unless the normal vectors are orthogonal. Consequently, the partition must consist of squares.

3.4.2 Analysis and Selection of Approximate Subspaces

The results of the previous section indicate the sensitivity of the performance of the element residual estimator to the partition geometry and even on the parity of the approximation when the local problems are solved exactly. In practice, it is necessary to formulate the element residual problems over a finite-dimensional subspace. The effect of solving the local residual problems *approximately* will now be studied. At first, one might expect that including the effects of solving approximately would only make a bad situation worse. As a matter of fact, a judiciously chosen approximate solution may actually *improve* matters.

3.4.2.1 Influence of Approximate Subspaces The *a posteriori* error estimate on element K is obtained by solving a local residual problem of the form

$$\phi_K \in Y_K : B_K(\phi_K, v) = F_K(v) - B_K(u_X, v) + \int_{\partial K} \left\langle \frac{\partial u_X}{\partial \boldsymbol{n}_K} \right\rangle v \, ds \quad \forall v \in Y_K \tag{3.68}$$

where Y_K is a finite-dimensional subspace of $H^1(K)/\mathbb{R}$. The choice of subspace for the error residual problem has a significant effect on the performance of the error estimator, as the following simple illustration shows. Consider the problem

$$-\Delta u = f \text{ in } \Omega = (0,1) \times (0,1). \tag{3.69}$$

The boundary conditions and data f are chosen so that the true solution is again, as in (3.59), given by

$$u(x,y) = x(x^p - 1) \tag{3.70}$$

for $p \in \mathbb{N}$. It is important to note that the solution and the data are regular, so that they are not the source of any strange behavior. The problem is solved using elements of degree p on the meshes consisting of lines parallel to the x and y axes. The finite element approximation is then identical to the interpolant at the Gauss–Lobatto points. The error residual problem will be solved using various choices of subspace:

Full Space

$$\widehat{Y} = \widehat{Q}(p+1)\backslash\mathbb{R}$$

Table 3.1 Performance of error estimators on uniform partition

p	$\|\|\|e\|\|\|_K^2$	Full	Uniform	Legendre	Lobatto
1	$\frac{1}{3}h^4$	$\frac{1}{3}h^4$	$\frac{1}{3}h^4$	$\frac{1}{3}h^4$	$\frac{1}{3}h^4$
2	$\frac{1}{20}h^6$	$\frac{3}{10}h^6$	$\frac{1}{20}h^6$	0	$\frac{1}{20}h^6$
3	$\frac{1}{175}h^8$	$\frac{1}{175}h^8$	$\frac{3565126227}{640000375000}h^8$	$\frac{1}{250}h^8$	$\frac{1}{175}h^8$
4	$\frac{1}{1764}h^{10}$	$\frac{5}{882}h^{10}$	$\frac{64}{132741}h^{10}$	0	$\frac{1}{1764}h^{10}$

Uniform Basis

$$\widehat{Y} = \text{span}\left\{\{1,\ldots,\widehat{x}^p\}\, W_{p+1}(\widehat{y}), W_{p+1}(\widehat{x})\, \{1,\ldots,\widehat{y}^p\}, W_{p+1}(\widehat{x})W_{p+1}(\widehat{y})\right\}$$

where

$$W_{p+1}(s) = \Pi_{j=0}^{p}(s - s_j)$$

and $s_j = -1 + 2j/p$, $j = 0, \ldots, p$.

Legendre Basis

$$\widehat{Y} = \text{span}\left\{\{1,\ldots,\widehat{x}^p\}\, P_{p+1}(\widehat{y}), P_{p+1}(\widehat{x})\, \{1,\ldots,\widehat{y}^p\}, P_{p+1}(\widehat{x})P_{p+1}(\widehat{y})\right\}$$

where P_{p+1} is the Legendre polynomial of degree $p + 1$.

Lobatto Basis

$$\widehat{Y} = \text{span}\left\{\{1,\ldots,\widehat{x}^p\}\, L_{p+1}(\widehat{y}), L_{p+1}(\widehat{x})\, \{1,\ldots,\widehat{y}^p\}, L_{p+1}(\widehat{x})L_{p+1}(\widehat{y})\right\}$$

where $L_{p+1}(s) = (1 - s^2)P'_{p+1}(s)$.

The subspace Y_K used for the solution of the residual problem is then

$$Y_K = \left\{\widehat{v} \circ F_K^{-1} : \widehat{v} \in \widehat{Y}\right\}, \tag{3.71}$$

where F_K is an affine mapping from the reference element onto K.

In each case, the error residual problem is solved and the error estimator computed. Table 3.1 shows the results obtained for uniform mesh spacing of size h.

Despite the fact that each of the subspaces is apparently reasonable and might be used for practical computation, it is observed that the performance of the error estimator is extremely sensitive to the choice of subspace used to solve the residual problem. In some cases the estimator is a gross overestimate,

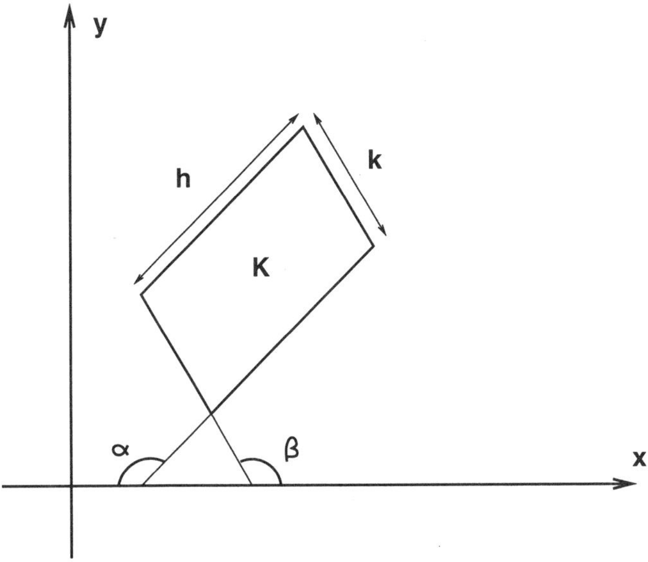

Fig. 3.1 Notation for parallelogram element.

yet in others the estimated error is zero, despite the true error being nonzero. Moreover, increasing the dimension of the subspace does not improve the performance of the error estimator; the full space is consistent only when the polynomial degree is odd, in agreement with the results of the previous section. The Lobatto basis is the only choice that is consistent in all cases.

It is worth emphasizing that this behavior has occurred for a very simple model problem. The purpose is to illustrate that even in such an innocuous, simplified setting the error estimator can give grossly misleading results inconsistent with the actual error unless the subspace used to approximate the residual problem is chosen with great care. In practical computations the true solution may be singular and the partition highly irregular, and one must anticipate these and further problems will arise.

We conclude this section with a positive result concerning the use of the Lobatto space.

Lobatto Subspace on Parallel Meshes Let $\mathcal{P}$ be a regular partitioning of the domain Ω and $F_K : \widehat{K} \mapsto K$ be an invertible, bilinear transformation of the reference element onto an element K. We shall consider the class of *parallel meshes*, whereby each element K is a parallelogram with sides of length h and k making angles α and β with the coordinate axes (see Figure 3.1). It is assumed that there exist fixed positive constants C and θ such that for all

$K \in \mathcal{P}$

$$\frac{1}{C} \leq \frac{h}{k} \leq C; \quad \theta \leq |\beta - \alpha| \leq \pi - \theta. \tag{3.72}$$

If h is the maximum element size, then the constants C and θ should be independent of h. Strictly speaking, parallel meshes are mild distortions of the partitions described above.

Theorem 3.6 *Suppose that the mesh is parallel and the true solution u belongs to $H^{p+2}(\Omega)$. If the error is properly $O(h^p)$ and $p > 1$, then the error estimator obtained using the Lobatto basis is asymptotically exact:*

$$\lim_{h \to 0} \frac{\eta}{|||e|||} = 1. \tag{3.73}$$

Proof. See reference [3]. ■

3.4.3 Conclusions

The error residual method for *a posteriori* error estimation is based on solving local residual problems for the error. The estimator is quite sensitive to the choice of subspace used to approximate the problem. In particular, using a full space of polynomials is inappropriate since the resulting estimator may give results completely inconsistent with the actual error. By considering appropriate subspaces of polynomials the performance of the estimator can be improved. Theorem 3.6 shows that if the mesh is parallel and the true solution smooth, then the resulting estimator is consistent with the actual error for degree $p > 1$ finite element approximation. However, the estimator is still inconsistent on nonuniform meshes even when $p = 1$.

The results can be explained as follows. The true solution of the residual problem can be regarded as being comprised of a number of components, some of which correspond to actual modes in the true error and other *spurious modes* that arise from the formulation of the problem using inexact boundary data. When a full space is used to approximate the problem, all of these components are present in the approximation. The resulting estimator is inconsistent owing to the contributions from the spurious modes. However, when a subspace Y_K is used to approximate the residual problem, then the components orthogonal to Y_K are not present in the approximation. If the subspace can be chosen so that it is precisely the spurious modes that are filtered out, then the consistency of the estimator will be recovered.

Thus, it is not surprising that the estimators are so sensitive to the choice of subspace. The results obtained using the subspaces in the illustration can be interpreted in the light of this explanation as follows. First, the subspace based on Uniform Nodes is inappropriate since it still contains some spurious modes when the degree p exceeds two, leading to overestimates of the error. On the opposite extreme, the subspace based on Legendre Nodes is inappropriate since it is not only orthogonal to the spurious modes but also orthogonal to modes that represent the actual error, leading to gross underestimates of the error. The subspace based on Lobatto Nodes is orthogonal to spurious modes but at the same time provides sufficient resolution of the true modes (whenever $p > 1$).

This state of affairs is somewhat unsatisfactory. However, useful estimators may be obtained, provided that the approximate subspace is selected carefully. An alternative possibility is to construct the boundary data for the underlying problem differently—for instance, using the equilibration procedures discussed in Chapter 6.

3.5 BIBLIOGRAPHICAL REMARKS

The subdomain residual method was devised by Babuška and Rheinboldt [20, 22, 23]. The approach is noteworthy because it provides the first general approach to *a posteriori* error estimation with firm theoretical foundations. In the present discussion, it was assumed that the partition $\mathcal{P}$ is regular, although extension to more general meshes is possible. In particular, the work of Babuška and Miller [19] allowed partitions with hanging nodes.

The bubble space for the element residual method applied to first-order finite elements on triangular elements suggested by Bank and Weiser, contains only edge bubble functions. The inclusion of an interior bubble function [109] simplifies the proof of the fact that the estimator provides two-sided bounds on the error, but is unnecessary, as was shown by Nochetto [84]. The element residual method for quadrilateral elements was presented in Demkowicz *et al.* [55, 56]

The discussion of the influence and selection of subspaces is based on references [2] and [3], where the asymptotic exactness on quadrilateral elements was also established. The asymptotic exactness of the Bank–Weiser estimator for piecewise affine finite element approximation on parallel meshes is given in reference [62].

4

Recovery-Based Error Estimators

The finite element method produces the optimal approximation from the finite element subspace. However, it is frequently the case that the finite element analyst is more interested in the gradient of the finite element approximation, than in the approximation itself. For instance, in computational elasticity, the stresses and strains are the primary concern, rather than the displacements. Furthermore, the normal component of the gradient of the approximation is generally discontinuous across the element boundaries, meaning that the practitioner is presented with a discontinuous approximation to the main quantity of interest. For this reason, many finite element packages incorporate a post-processing procedure whereby the discontinuous approximation to the gradient is smoothed before being presented to the user.

The reasons for performing such a post-processing are not purely cosmetic: Under certain circumstances it is found that the accuracy of the smoothed gradient is superior to the approximation provided by the untreated gradient of the original finite element approximation. A rather natural approach to *a posteriori* error estimation is based on measuring the difference between the direct and post-processed approximations to the gradient. This ostensibly rather crude approach can result in astonishingly good estimates for the true error.

In order to describe this procedure more formally, consider the model problem in Section 1.4 and suppose that the absolute term c vanishes, so that the true error measured in the energy norm is

$$|||e|||^2 = \int_\Omega |\nabla u - \nabla u_X|^2 \, \mathrm{d}x. \tag{4.1}$$

Of course, if the true gradient were in hand, then it would be a relatively easy matter to substitute it into this expression and to calculate the true error exactly. Intuitively, a reasonable error estimator should be obtained by using a suitable approximation to the gradient in place of ∇u. In particular, let $G_X[u_X]$ denote an approximation to the gradient of the true solution obtained by a suitable post-processing of the finite element approximation. The *a posteriori* error estimator is then simply taken as

$$\eta^2 = \int_\Omega |G_X[u_X] - \nabla u_X|^2 \, \mathrm{d}x. \tag{4.2}$$

The case $c \not\equiv 0$ is dealt with by arguing that the dominant term in the error is the component containing the derivatives, and so it should be enough to estimate this dominant part only. In effect, the absolute term is simply ignored.

The appeal of such procedures to practitioners is easy to appreciate. The ready availability of a post-processed gradient in the finite element package means that the estimator may be easily implemented. This approach allows considerable leeway in the selection of the post-processed gradient, and there are as many different estimators as there are post-processing techniques.

One class of techniques that is particularly popular with the engineering community is averaging methods. The gradient of the finite element approximation provides a discontinuous approximation to the true gradient. This may be used to construct an approximation at each node by averaging contributions from each of the elements surrounding the node. These values may then be interpolated to obtain a continuous approximation over the whole domain. The specific steps used to construct the averaged gradient at the nodes distinguish the various estimators and have a major influence on the accuracy and robustness of the ensuing estimator. The present chapter is concerned with presenting a fairly general framework to assist in the development and analysis of such estimators. One fringe benefit is that it will be found that some rigorously analyzed estimators obtained in quite different ways fall within the framework of corresponding to a particular choice of recovered gradient G_X.

4.1 EXAMPLES OF RECOVERY-BASED ESTIMATORS

This section illustrates the ideas discussed in the previous section by deriving error estimators for some particular types of finite element approximation scheme.

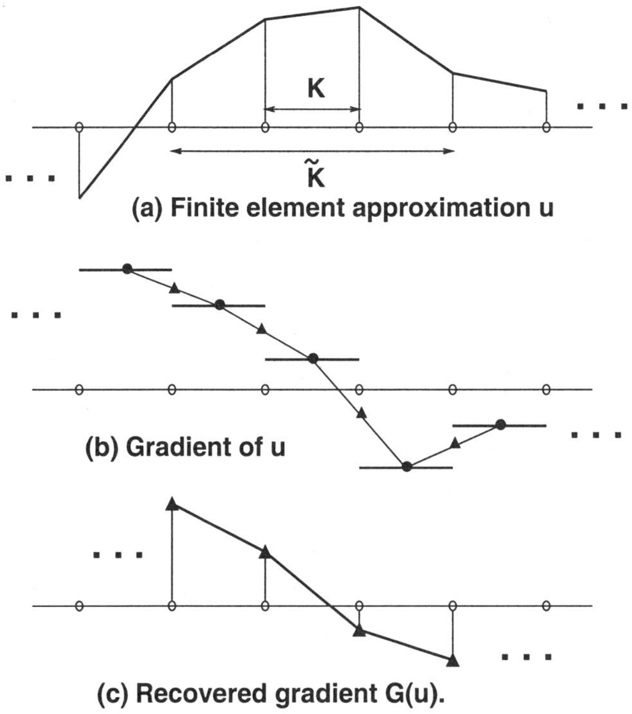

Fig. 4.1 Construction of recovery operator $\boldsymbol{G}_X$ from piecewise linear approximation in one dimension.

4.1.1 An Error Estimator for a Model Problem in One Dimension

Consider piecewise linear approximation of the simple one-dimensional problem

$$-u'' = f \text{ on } (0,1); \quad u(0) = u(1) = 0. \tag{4.3}$$

There are many types of *a posteriori* error estimator available for this situation. The purpose here is not to obtain new results, but to show how an existing estimator fits within the framework to be developed.

The first step is to define the procedure for smoothing the gradient of the finite element approximation. The procedure is shown graphically in Figure 4.1. The post-processed, or recovered, gradient is denoted by $\boldsymbol{G}_X(u_X)$, where u_X is the finite element approximation. The recovered gradient is itself piecewise linear, with values at the nodes obtained by first interpolating the gradient of the finite element approximation at the centroids of the elements sharing the node. The estimator associated with element K is then defined to be

$$\eta_K = \|\boldsymbol{G}_X(u_X) - u_X'\|_{L_2(K)} \,. \tag{4.4}$$

This quantity is computable once the finite element solution is in hand. The estimator is precisely the estimator originally proposed and analyzed by Babuška and Rheinboldt [21, Definition 6.3]. Their derivation of the estimator was based on an entirely different argument involving the local projection of the error onto a quadratic bubble function that vanishes at the nodes. For further details, the reader is referred to reference [21], where numerical examples illustrating the effectiveness of this estimator will also be found.

The main ingredient in deriving the estimator centers on the construction of the recovery operator $\boldsymbol{G}_X$. It is worth considering the steps involved in more detail. First, the reason behind seeking a recovered gradient in the form of a piecewise linear function is mainly one of convenience. Specifically, the finite element approximation is itself piecewise linear, meaning that the same numerical procedures already present in the finite element code may be reused to store and handle the post-processed gradient.

The reason for sampling the gradient at the centroid of the elements is related to the well known fact [115] that the gradient at the centroids is *superconvergent*. It is natural to utilize the gradient sampled at these points in seeking to produce an accurate post-processed gradient.

The relevant superconvergence result for this situation is known from the work of Zlamal [124]: If $u \in H^3(\Omega)$ and the partition is uniform, then

$$|u_X - I_X u|_{H^1(\Omega)} \leq Ch^2 \, |u|_{H^3(\Omega)} \tag{4.5}$$

where I_X is the X-interpolant. The accuracy of piecewise linear finite elements measured in the energy norm is known to be $\mathcal{O}(h)$ (unless the solution is trivial). Consequently, the estimate (4.5) means that the finite element solution is a better approximation to the nodal interpolant of the true solution than it is to the true solution itself. Suppose for the moment that the true solution u happens to be a quadratic polynomial (or at least, may be well-approximated by a quadratic polynomial). It is a simple exercise to verify that the recovery operator described above satisfies

$$\boldsymbol{G}_X(I_X v) \equiv I_X(v') \equiv v' \quad \forall v \in \mathbb{P}_2. \tag{4.6}$$

For instance, first observe that if v is quadratic, then the value of $(I_X v)$ sampled at the element centroids coincides with the true gradient v' at the centroids. Consequently, since the recovered values of the gradients at the element vertices are obtained by linear interpolation of the values at the centroids, it follows that the values will coincide with the value of the actual gradient at the vertices.

The property (4.6), coupled with the knowledge that the interpolant of u is superconvergent to the finite element solution, leads one to conjecture that when the recovery operator is applied to the finite element solution, then the approximation

$$\boldsymbol{G}_X(u_X) \approx u' \tag{4.7}$$

is itself superconvergent. This result will be formulated more precisely and proved later.

The interpretation of the recovery procedure as a linear operator will prove useful in the later analysis. For this example, the operator $\boldsymbol{G}_X$ is linear and has values on the element K that are determined completely by local information based on values of the direct approximation to the gradient sampled on the set $\widetilde{K}$ as shown in Figure 4.1. The operator is linear and bounded in the sense that

$$|\boldsymbol{G}_X(v)|_{L_\infty(K)} \leq 3\,|v|_{W^{1,\infty}(\widetilde{K})}\,. \tag{4.8}$$

4.1.2 An Error Estimator for Bilinear Finite Element Approximation

Consider the finite element approximation of Poisson's equations using piecewise bilinear approximation in two dimensions. For the sake of simplicity assume that each of the elements K is a square with sides of length h parallel to the axes.

By analogy with the previous one-dimensional example, a post-processing procedure is defined by sampling the gradients of the finite element approximation at the centroids of the elements. These values are then averaged to produce an approximation to the gradient at the nodes, as shown in Figure 4.2. As before, the recovered gradient $\boldsymbol{G}_X(u_X)$ is selected to be the bilinear function in each component that interpolates the values recovered at the nodes. The estimator η_K on element K is

$$\eta_K = \|\boldsymbol{G}_X(u_X) - \boldsymbol{\nabla} u_X\|_{L_2(K)}\,. \tag{4.9}$$

As before, the reason behind sampling at the centroids is founded on the superconvergence property demonstrated by Zlamal [124, 125]:

$$|u_X - I_X u|_{H^1(\Omega)} \leq Ch^2\,|u|_{H^3(\Omega)}\,, \tag{4.10}$$

where I_X is the bilinear interpolant at the vertices of the partition. The operator $\boldsymbol{G}_X$ is linear and bounded since

$$|\boldsymbol{G}_X[v]|_{L_\infty(K)} \leq C\,|v|_{W^{1,\infty}(\widetilde{K})} \quad \forall v \in X. \tag{4.11}$$

Furthermore, it is found that

$$\boldsymbol{G}_X[I_X v] \equiv I_X(\boldsymbol{\nabla} v) \tag{4.12}$$

whenever $v \in \mathbb{P}_2$. This condition holds trivially when $v \in \mathbb{Q}_1$, so it remains to consider the cases when v is x^2 or y^2. By symmetry, it suffices to consider the first case when $v = x^2$. First, consider the single element $(0,h) \times (0,h)$. The interpolant $I_X v$ is given by hx and therefore, at the element centroid,

$$\boldsymbol{\nabla}(I_X v) = \begin{bmatrix} h \\ 0 \end{bmatrix}.$$

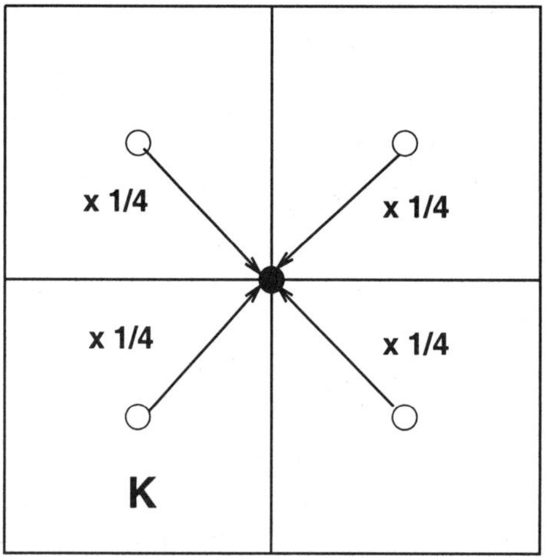

(a) Scheme for new estimator.

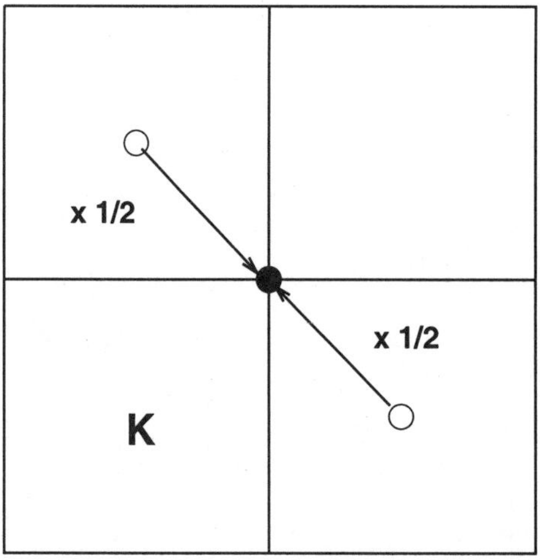

(b) Scheme for Kelly et al. estimator.

Fig. 4.2 Construction of recovered gradient at vertex of element K. The value at • is a linear combination of the values at ∘ using the weights indicated.

The value of $I_X \nabla v$ at the centroid is the average of the values of the gradients at the element vertices, so that

$$I_X \nabla v = \begin{bmatrix} h \\ 0 \end{bmatrix}.$$

These arguments show that the values sampled at the centroids coincide. It follows that the values at the vertices will also coincide since the simple averaging is consistent with applying bilinear interpolation. Finally, since the values at the vertices coincide, it follows that (4.12) holds for the case when $v = x^2$.

It is interesting to compare the estimator associated with this recovery procedure with an estimator proposed by Kelly *et al.* [74]:

$$\hat{\eta}_K^2 = \frac{h}{24} \int_{\partial K} \left[\frac{\partial u_X}{\partial n} \right]^2 \mathrm{d}s, \tag{4.13}$$

where

$$\left[\frac{\partial u_X}{\partial n} \right] = \left. \frac{\partial u_X}{\partial n_K} \right|_K + \left. \frac{\partial u_X}{\partial n_{K'}} \right|_{K'} \tag{4.14}$$

is the familiar jump discontinuity in normal gradient of the finite element approximation on the edge between neighboring elements K and K'. Using the midpoint rule for integration along each side of the element, the estimator (4.13) may be rewritten as

$$\hat{\eta}_K^2 = \frac{h^2}{24} \sum_{\gamma \subset \partial K} \left[\frac{\partial u_X}{\partial n} \right]^2 \tag{4.15}$$

where the discontinuities are evaluated at the midpoints of the sides.

This estimator may be viewed as a modified recovery operator by taking the values of the recovered gradient at the vertices to be (see Figure 4.2)

$$\boldsymbol{G}_X[v](x, y) = \frac{1}{2} \left[\nabla v|_{x - \frac{1}{2}h, y + \frac{1}{2}h} + \nabla v|_{x + \frac{1}{2}h, y - \frac{1}{2}h} \right]. \tag{4.16}$$

The new recovery operator associated with this procedure gives rise to an estimator *identical* to $\hat{\eta}_K$. The estimator derived by Kelly *et al.* may therefore be viewed using the above framework. This approach leads straightforwardly to the result (4.15).

The recovery-based estimator (4.9), like (4.15), is found, after a lengthy but otherwise straightforward manipulation, to depend on the discontinuities in the finite element approximation to the gradient. However, the dependence is more intricate than in (4.15) involving, in addition, differences in tangential components at the centroids. The estimator (4.13) uses gradients sampled from the element K and elements sharing a common edge, while the estimator (4.15) also involves elements sharing a common node as shown in Figure 4.2.

It might seem that the estimator η_K is too complicated to be of practical use. However, it is in many ways much simpler than the Kelly estimator. For instance, with the Kelly estimator it is not obvious how one should define the value of the jump along the exterior boundary $\partial\Omega$. This difficulty does not arise with the recovery-based estimator. The above approach even provides the answer: The value should be taken to be the jump on the opposite side of the element.

4.2 RECOVERY OPERATORS

The purpose of this section is to formalize the essential ideas underlying the recovery procedures described in the previous section. In particular, we shall identify a set of conditions sufficient for the operators $\boldsymbol{G}_X$ guaranteeing that $\boldsymbol{G}_X(I_X u)$ is a good approximation to the true gradient ∇u.

Consistency Condition As with any numerical procedure, a basic requirement for the post-processing operator should be that in favorable circumstances the operator should reproduce the true gradient exactly. This type of condition is generally referred to as *consistency* in the context of a numerical algorithm. One consequence of the consistency condition is that if the finite element approximation is exact, then the true error and the estimate resulting from the recovery-based procedure will both vanish.

The consistency condition for the recovery operators is formulated as follows:

(R1) If u is a polynomial of degree $p+1$ on the patch $\widetilde{K}$ associated with an element K—that is, $u \in \mathbb{P}_{p+1}(\widetilde{K})$—then

$$\boldsymbol{G}_X(I_X u) = I_X \boldsymbol{\nabla} u \text{ on } K, \tag{4.17}$$

where I_X is the X-interpolant.

Observe that the polynomial space $\mathbb{P}_{p+1}$ is used for both triangular and quadrilateral elements.

Of course, the consistency condition does not determine the recovery operator $\boldsymbol{G}_X$ uniquely. The examples in the previous section show that it provides a natural and convenient criterion.

Localization Condition An important practical requirement is that $\boldsymbol{G}_X$ should be inexpensive to compute. Ideally, it should be possible to compute the recovered gradient $\boldsymbol{G}_X$ without recourse to global computations: Otherwise it would be simpler to resolve the original finite element problem on a finer partition. The most convenient schemes are those whereby the recovered gradient at a point $\boldsymbol{x}_0$ is a linear combination of values of the gradient of the finite element approximation sampled in a neighborhood of the point $\boldsymbol{x}_0$. Let

$\widetilde{K}$ denote the patch consisting of the element K and its neighboring elements, as in (1.30). The localization condition is as follows:

(R2) If $\boldsymbol{x}_0 \in K$, then the value of the recovered gradient $\boldsymbol{G}_X[v](\boldsymbol{x}_0)$ depends only on values of $\boldsymbol{\nabla} v$ sampled on the patch $\widetilde{K}$.

The condition is formulated over the patches $\widetilde{K}$ primarily for convenience. In principle, there is no reason why a larger patch should not be employed if desired. However, the recovery procedures described in the previous section indicate that the patch $\widetilde{K}$ provides a sufficiently flexible framework to incorporate most practical schemes.

Boundedness and Linearity Conditions The form of the recovered gradient should, ideally, be one that may be handled efficiently by the existing data structures within the finite element code. In particular, $\boldsymbol{G}_X$ should be a simple function that may be evaluated and integrated easily. If $\boldsymbol{G}_X$ is similar to functions belonging to the space used to construct the finite element subspace, then existing routines from the finite element code may be used to manipulate $\boldsymbol{G}_X$. These considerations lead to the following condition:

(R3) $\boldsymbol{G}_X : X \to X \times X$ is a linear operator, and there exists a constant C (independent of h) such that

$$\|\boldsymbol{G}_X[v]\|_{L_\infty(K)} \leq C\, |v|_{W^{1,\infty}(\widetilde{K})} \quad \forall K \in \mathcal{P} \quad \forall v \in X \tag{4.18}$$

where X is the finite element subspace.

4.2.1 Approximation Properties of Recovery Operators

The conditions (R1)–(R3) imply that the operator $\boldsymbol{G}_X$ possesses certain approximation properties. In particular, when u is smooth, $\boldsymbol{G}_X(I_X u)$ is a good approximation to $\boldsymbol{\nabla} u$. This notion is quantified in the following result, where the notation is described in Section 1.3.

Theorem 4.1 *Let $\mathcal{P}$ be a regular partitioning of the domain Ω into triangular or quadrilateral elements and let X be the finite element subspace based on polynomials of degree p. Suppose that $\boldsymbol{G}_X$ satisfies (R1)–(R3) and that $u \in H^{p+2}(\widetilde{K})$. Then,*

$$\|\boldsymbol{\nabla} u - \boldsymbol{G}_X(I_X u)\|_{L_2(K)} \leq C h_K^{p+1}\, |u|_{H^{p+2}(\widetilde{K})} \tag{4.19}$$

where $C > 0$ is independent of h_K and u.

Proof. Let $K \in \mathcal{P}$ be any element and, for the duration of the proof, write h in place of h_K. Suppose $u \in H^{p+2}(\widetilde{K})$ and let

$$E[u] = \boldsymbol{\nabla} u - \boldsymbol{G}_X\,(I_X u)$$

be the error in the recovered gradient. Let $w \in \mathbb{P}_{p+1}$ be a function to be selected later, and write

$$E[u] = \nabla(u-w) + (\nabla w - I_X \nabla w) + (I_X \nabla w - \boldsymbol{G}_X(I_X u))\,.$$

Invoking the consistency condition (R1) and the linearity condition (R3) enables the final term to be rewritten as

$$\boldsymbol{G}_X(I_X w) - \boldsymbol{G}_X(I_X u) = \boldsymbol{G}_X(I_X(w-u));$$

consequently, applying the triangle inequality leads to the conclusion

$$\begin{aligned} \|E[u]\|_{L_2(K)} &\leq \|\nabla(u-w)\|_{L_2(K)} + \|\nabla w - I_X \nabla w\|_{L_2(K)} \\ &\quad + \|\boldsymbol{G}_X(I_X(w-u))\|_{L_2(K)}\,. \end{aligned} \tag{4.20}$$

These terms are estimated in turn as follows: Firstly, it is clear that

$$\|\nabla(u-w)\|_{L_2(K)} \leq |u-w|_{H^1(\widetilde{K})}\,.$$

The approximation property of the X-interpolant given in Theorem 1.5 implies that

$$\|\nabla w - I_X \nabla w\|_{L_2(K)} \leq Ch^{p+1}\, |\nabla w|_{H^{p+1}(K)} \leq Ch^{p+1}\, |w|_{H^{p+2}(\widetilde{K})}\,,$$

and then by the triangle inequality we obtain

$$|w|_{H^{p+2}(\widetilde{K})} \leq |u|_{H^{p+2}(\widetilde{K})} + |u-w|_{H^{p+2}(\widetilde{K})}\,.$$

Finally, by Hölder's inequality we have

$$\|\boldsymbol{G}_X(I_X(w-u))\|_{L_2(K)} \leq Ch\, \|\boldsymbol{G}_X(I_X(w-u))\|_{L_\infty(K)}\,,$$

and then, thanks to the boundedness condition (R3) we obtain

$$\|\boldsymbol{G}_X(I_X(w-u))\|_{L_\infty(K)} \leq C\, |I_X(w-u)|_{W^{1,\infty}(\widetilde{K})}\,.$$

The patch $\widetilde{K}$ comprises of elements that are quasi-uniform of size h. Therefore, in view of the inverse estimate we have

$$|I_X(w-u)|_{W^{1,\infty}(\widetilde{K})} \leq Ch^{-1}\, \|I_X(w-u)\|_{L_\infty(\widetilde{K})}\,.$$

By the equivalence of norms on the finite-dimensional space X and a scaling argument, it follows that

$$\|I_X(w-u)\|_{L_\infty(\widetilde{K})} \leq Ch^{-1}\, \|I_X(w-u)\|_{L_2(\widetilde{K})}\,.$$

The triangle inequality shows that

$$\|I_X(w-u)\|_{L_2(\widetilde{K})} \leq \|w-u\|_{L_2(\widetilde{K})} + \|(w-u) - I_X(w-u)\|_{L_2(\widetilde{K})}$$

and applying the approximation property of the X-interpolant results in the estimate

$$\|I_X(w-u)\|_{L_2(\tilde{K})} \leq \|w-u\|_{L_2(\tilde{K})} + Ch^{p+1} |w-u|_{H^{p+1}(\tilde{K})} .$$

With the aid of Theorem 1.1 and taking the infimum over all functions $w \in \mathbb{P}_{p+1}$, we conclude that:

$$\|\nabla(u-w)\|_{L_2(K)} \leq Ch^{p+1} |u|_{H^{p+2}(\tilde{K})}$$

$$\|\nabla w - I_X \nabla w\|_{L_2(K)} \leq Ch^{p+1} |u|_{H^{p+2}(\tilde{K})}$$

and

$$\|G_X(I_X(w-u))\|_{L_2(K)} \leq Ch^{p+1} |u|_{H^{p+2}(\tilde{K})} .$$

The desired result follows at once from (4.20). ■

This local result can be used to obtain a global estimate:

Corollary 4.2 *Suppose that the hypotheses of Theorem 4.1 hold. Then,*

$$\|\nabla u - G_X(I_X u)\|_{L_2(\Omega)} \leq Ch^{p+1} |u|_{H^{p+2}(\Omega)} \tag{4.21}$$

where $C > 0$ *is independent of* h, *where* $h = \max h_K$.

Proof. Sum over the elements using Theorem 4.1. ■

4.3 THE SUPERCONVERGENCE PROPERTY

It has been shown that if a recovery operator G_X is found satisfying the conditions (R1)–(R3), then applying the operator to $I_X u$ furnishes good approximations to the derivatives of u. Of course, in practice the interpolant $I_X u$ is not available, and we therefore now consider the effect of applying G_X to the finite element approximation itself. In some circumstances, for instance if the *superconvergence phenomenon* is present, then applying G_X to the finite element approximation itself gives correspondingly good approximations to the derivatives.

The superconvergence property may be understood by recalling that the classic *a priori* estimate for the error in the finite element approximation takes the form

$$|||u - u_X||| \leq Ch^{\nu} \|u\|_{H^{p+\tau}(\Omega)} \tag{4.22}$$

for appropriate positive constants ν and τ. The estimate (4.22) is optimal in the sense that the exponent of h is the largest possible. In fact, for the h-version finite element method one may show that provided that the pth-order derivatives of the true solution u do not vanish identically over the whole domain, then

$$|||e||| \geq C(u)h^{p} \tag{4.23}$$

for some positive constant $C(u)$ depending only on u.

Superconvergence is said to be present if, under appropriate regularity conditions on the partition and the true solution, an estimate of the form

$$|u_X - I_X u|_{H^1(\Omega)} \le C(u) h^{p+1} \tag{4.24}$$

holds. Comparing (4.23) and (4.24) shows that ∇u_X is a better approximation to $\nabla I_X u$ than it is to ∇u. The superconvergence property is embodied in the assumption:

(SC) There exist positive constants C and $\tau \in (0, 1]$ that are independent of h such that

$$|u_X - I_X u|_{H^1(\Omega)} \le C(u) h^{p+\tau}. \tag{4.25}$$

The precise assumptions used to obtain such estimates differ according to the type of finite element approximation scheme being used. *It should be stressed that superconvergence only occurs in very special circumstances.* A survey of superconvergence results is contained in Križek and Neitaanmaki [76].

The next result confirms the intuitive expectation concerning the accuracy of the recovery operator applied to the finite element approximation directly in the presence of superconvergence.

Lemma 4.3 *Suppose $u \in H^{p+2}(\Omega)$, $\boldsymbol{G}_X$ satisfies (R1)–(R3), and (SC) holds. Then*

$$\|\boldsymbol{\nabla} u - \boldsymbol{G}_X(u_X)\|_{L_2(\Omega)} \le C(u) h^{p+\tau} \tag{4.26}$$

holds where $C > 0$ is independent of h.

Proof. By the triangle inequality and the linearity of $\boldsymbol{G}_X$, we obtain

$$\|\boldsymbol{\nabla} u - \boldsymbol{G}_X(u_X)\|_{L_2(K)} \le |\boldsymbol{\nabla} u - \boldsymbol{G}_X(I_X u)|_{L_2(K)} + \|\boldsymbol{G}_X(I_X u - u_X)\|_{L_2(K)} .$$

The boundedness property (R3) of $\boldsymbol{G}_X$ implies that, with the aid of an inverse estimate,

$$\|\boldsymbol{G}_X(I_X u - u_X)\|_{L_2(K)} \le C \, \|I_X u - u_X\|_{H^1(\tilde{K})} .$$

Summing over the elements and using Theorem 4.1 and the superconvergence property (SC) gives the result as claimed. ∎

4.4 APPLICATION TO A POSTERIORI ERROR ESTIMATION

Consider the class of error estimators obtained by using $\boldsymbol{G}_X(u_X)$ instead of $\boldsymbol{\nabla} u$ in the expression for the error. That is, the estimator on element K is η_K, where

$$\eta_K = \|\boldsymbol{G}_X(u_X) - \boldsymbol{\nabla} u_X\|_{L_2(K)} . \tag{4.27}$$

The global estimator is obtained by summing contributions from the elements. The next result shows that a recovery-based estimator will give an asymptotically exact estimate of the true error if the superconvergence property is present.

Theorem 4.4 *Let η be the* a posteriori *error estimator defined above. Assume that (SC), (R1)–(R3), and (4.23) hold. Then*

$$\lim_{h \to 0} \frac{\eta}{|||e|||} = 1. \tag{4.28}$$

Proof. By the triangle inequality and the foregoing results, we obtain

$$\begin{aligned} |\eta - |||e|||\,| &\leq \|\boldsymbol{G}_X(u_X) - \nabla u_X - \nabla e\|_{L_2(\Omega)} + \|c\|_{L_\infty(\Omega)} \|e\|_{L_2(\Omega)} \\ &= \|\boldsymbol{G}_X(u_X) - \nabla u\|_{L_2(\Omega)} + \|c\|_{L_\infty(\Omega)} h^\mu\, |||e||| \\ &\leq C(u) h^{p+\tau} + \|c\|_{L_\infty(\Omega)} h^\mu\, |||e|||, \end{aligned}$$

since

$$\|e\|_{L_2(\Omega)} \leq C h^\mu\, |||e|||$$

for suitable positive constants C and μ depending on the smoothness of the solution and the domain Ω. The result then follows from (4.23). ∎

Theorem 4.4 reduces the problem of finding *a posteriori* error estimators to using the existing superconvergence results to define an appropriate recovery operator $\boldsymbol{G}_X$. Consequently, whenever we have superconvergence results for a particular finite element scheme, it is then possible to define an *a posteriori* error estimator that is asymptotically exact. The examples from earlier in the chapter fall within the framework covered by Theorem 4.4, and as an immediate consequence they are asymptotically exact provided that the conditions are favorable for the superconvergence property to hold.

As mentioned before, the superconvergence property is quite fragile and depends on rather stringent assumptions concerning the regularity of the true solution and the partition, which are frequently not satisfied in practical computation, particularly in the case of adaptive refinement procedures where the highly localized nature of the refinements invalidates many of the superconvergence results. Nevertheless, it is found that recovery-based estimators, while no longer being asymptotically exact, continue to perform astonishingly well in such situations. The reason behind the robustness of such estimators is not fully understood at present.

4.5 CONSTRUCTION OF RECOVERY OPERATORS

The framework for constructing recovery operators embodied in conditions (R1)–(R3) is fairly prescriptive in determining the possible recovery operators.

The next result provides a more precise characterization of what freedom is left in selecting a recovery operator.

Lemma 4.5 *Suppose the recovery operator $\boldsymbol{G}_X$ satisfies conditions (R1)–(R3). Then $\boldsymbol{G}_X$ is of the form*

$$\boldsymbol{G}_X[v] = \sum_{k \in \mathcal{N}} \boldsymbol{g}_k[v]\theta_k, \tag{4.29}$$

where $\{\theta_k : k \in \mathcal{N}\}$ is the Lagrange basis associated with the nodes $\{\boldsymbol{x}_k : k \in \mathcal{N}\}$; and $\boldsymbol{g}_k$ are linear functionals that are bounded in the sense that there exists a constant C such that for all $v \in W^{1,\infty}(\widetilde{\Omega}_k)$,

$$|\boldsymbol{g}_k[v]| \leq C\,|v|_{W^{1,\infty}(\widetilde{\Omega}_k)} \tag{4.30}$$

and, for all polynomials $v \in \mathbb{P}_{p+1}$,

$$\boldsymbol{g}_k[I_X v] = \boldsymbol{\nabla} v(\boldsymbol{x}_k) \tag{4.31}$$

where $\widetilde{\Omega}_k$ is the patch of elements containing the node $\boldsymbol{x}_k$ defined by

$$\widetilde{\Omega}_k = \left\{K : \boldsymbol{x}_k \in \overline{K}, K \in \mathcal{P}\right\}. \tag{4.32}$$

Proof. The condition (R3) requires that $\boldsymbol{G}_X \in X \times X$ and it follows that we may write

$$\boldsymbol{G}_X[v] = \sum_{k \in \mathcal{N}} \boldsymbol{g}_k[v]\theta_k$$

where $\boldsymbol{g}_k[v]$ are *linear* functionals, again thanks to condition (R3). The property characterizing the Lagrange basis functions means that

$$\boldsymbol{g}_k[v] = \boldsymbol{G}_X[v](\boldsymbol{x}_k) \quad k \in \mathcal{N}.$$

Furthermore, due to the consistency condition (R1), it follows that if v is a polynomial of degree $p+1$, then

$$\boldsymbol{G}_X[I_X v](\boldsymbol{x}_k) = I_X\left(\boldsymbol{\nabla} v\right)(\boldsymbol{x}_k) = \boldsymbol{\nabla} v(\boldsymbol{x}_k)$$

and hence we conclude that

$$\boldsymbol{g}_k[I_X v] = \boldsymbol{\nabla} v(\boldsymbol{x}_k)$$

whenever v is a polynomial of degree $p+1$.

The localization condition (R2) means that $\boldsymbol{G}_X[v](\boldsymbol{x}_k)$ depends only on the values of the gradient $\boldsymbol{\nabla} v$ sampled on the patch $\widetilde{\Omega}_k$ defined in the statement of the result. Finally, condition (R3) means that (4.30) must hold. ■

4.6 THE ZIENKIEWICZ–ZHU PATCH RECOVERY TECHNIQUE

Lemma 4.5 shows that the process of constructing a recovery operator satisfying conditions (R1)–(R3) reduces to selecting a post-processing procedure for the values of the gradients at the nodes that satisfies conditions (4.30)–(4.31).

The purpose of the present section is to describe a particular procedure for post-processing the finite element approximation to obtain values of the fluxes at the nodes. These values are then used to obtain the globally reconstructed fluxes as described in Lemma 4.5 and, finally, the recovered flux $\boldsymbol{G}_X(u_X)$ is used to produce an error estimator as in Section 4.4.

The so-called *superconvergent patch recovery (SPR)* procedure was developed by Zienkiewicz and Zhu [122, 123]. The basic method is applicable to partitions consisting of triangular or quadrilateral elements. The procedure will be explained by considering piecewise linear and quadratic finite element approximation on triangular elements.

4.6.1 Linear Approximation on Triangular Elements

Every node $\boldsymbol{x}_k$, $k \in \mathcal{N}$, is an element vertex, and consequently, the patch $\widetilde{\Omega}_k$ consists of the elements having $\boldsymbol{x}_k$ as a vertex. The gradient is sampled at the centroid $\boldsymbol{c}_K$ of each element K in the patch $\widetilde{\Omega}_k$, as illustrated in Figure 4.3(a).

The values of the gradients sampled at the centroids in the patch are used to produce a recovered value at the central node by first fitting each component of the flux data to a function of the form

$$\boldsymbol{p}(\boldsymbol{x})^\top \boldsymbol{\alpha} \tag{4.33}$$

where the vector $\boldsymbol{p}$ is based on the polynomial space $\mathbb{P}_1$ used in the construction of the finite element space X,

$$\boldsymbol{p}(\boldsymbol{x}) = \begin{bmatrix} 1 \\ x \\ y \end{bmatrix} \tag{4.34}$$

and $\boldsymbol{\alpha}$ is a vector of coefficients to be determined. The coefficients are determined by a discrete least squares fit based on the values at the sampling points: Thus $\boldsymbol{\alpha}$ is the unique minimizer of (in the case of the x-component of the gradient),

$$\mathcal{J}(\boldsymbol{\alpha}) = \sum_{K \subset \widetilde{\Omega}_k} \left(\frac{\partial u_X}{\partial x}(\boldsymbol{c}_K) - \boldsymbol{p}(\boldsymbol{c}_K)^\top \boldsymbol{\alpha} \right)^2 . \tag{4.35}$$

The Euler condition for the minimizer reveals that $\boldsymbol{\alpha}$ is the solution of the matrix equation

$$\boldsymbol{M}\boldsymbol{\alpha} = \boldsymbol{b} \tag{4.36}$$

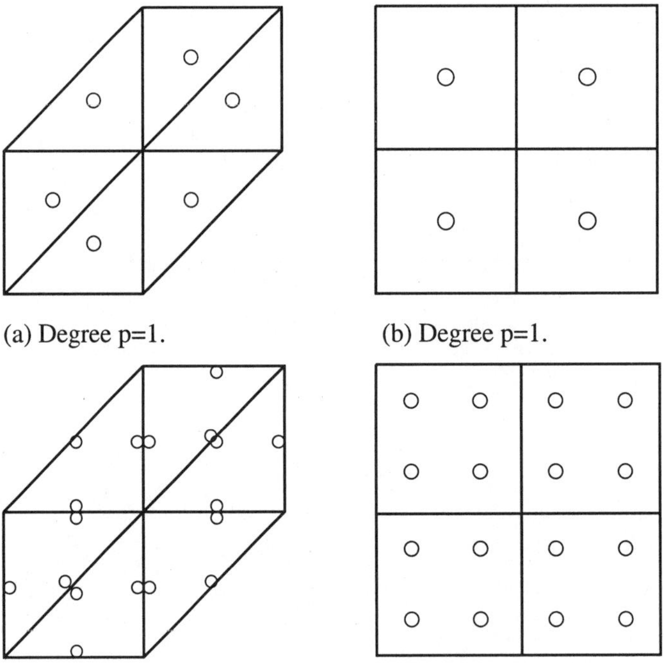

(a) Degree p=1. (b) Degree p=1.

(c) Degree p=2. (d) Degree p=2.

Fig. 4.3 Sampling points employed in the superconvergent patch recovery technique.

where $\boldsymbol{M}$ is the matrix

$$\boldsymbol{M} = \sum_{K \subset \tilde{\Omega}_k} \boldsymbol{p}(\boldsymbol{c}_K)\boldsymbol{p}(\boldsymbol{c}_K)^\top \tag{4.37}$$

and $\boldsymbol{b}$ is the vector

$$\boldsymbol{b} = \sum_{K \subset \tilde{\Omega}_k} \boldsymbol{p}(\boldsymbol{c}_K)^\top \frac{\partial u_X}{\partial x}(\boldsymbol{c}_K). \tag{4.38}$$

Finally, the recovered (x-component) of the flux at the central node is defined to be

$$g_k[u_X] = \boldsymbol{p}(\boldsymbol{x}_k)^\top \boldsymbol{\alpha}. \tag{4.39}$$

For the very regular, parallel mesh shown in Figure 4.3, it is known [80] that this recovered value satisfies the conditions of Lemma 4.5.

4.6.2 Quadratic Approximation on Triangular Elements

The patch recovery procedure is modified for the piecewise quadratic approximation on triangles as follows.

The first difference is the sampling points for the gradients are modified so that the gradient is sampled at the mid-side points as shown in Figure 4.3(c). Furthermore, the form of the function used to fit the gradient is modified to be

$$\boldsymbol{p}(\boldsymbol{x}) = \begin{bmatrix} 1 \\ x \\ y \\ x^2 \\ xy \\ y^2 \end{bmatrix}. \tag{4.40}$$

The post-processing of the gradients at the nodes located at the *vertices* is different from the post-processing used to construct the values at the nodes located at the *mid-side* points.

The values at the vertices are constructed using the same procedure as for the case of linear finite element approximation. Namely, the matrix equation (4.33) is assembled and solved for each of the vertex nodes. Of course, the definitions of the matrix $\boldsymbol{M}$ and vector $\boldsymbol{b}$ are modified to reflect the different choice of functions used in $\boldsymbol{p}$ and the different set of sampling points, but the general form is the same. As before, the recovered value of the gradient at the vertex is obtained by evaluating the fitted data as in (4.39),

$$g_k[v] = \boldsymbol{p}(\boldsymbol{x}_k)^\top \boldsymbol{\alpha}. \tag{4.41}$$

The recovered value at a mid-side point $\boldsymbol{x}_k$ makes use of the fitted data using the patches associated with each of the vertices $\boldsymbol{x}_\ell$ and $\boldsymbol{x}_r$ of the edge where the mid-side point is located. Thus, if coefficients $\boldsymbol{\alpha}_\ell$ and $\boldsymbol{\alpha}_r$ are obtained when fitting the data at the endpoints, then the recovered value at the mid-side node is taken to be a simple arithmetic average of the values

$$g_k[v] = \frac{1}{2}\boldsymbol{p}(\boldsymbol{x}_k)^\top \boldsymbol{\alpha}_\ell + \frac{1}{2}\boldsymbol{p}(\boldsymbol{x}_k)^\top \boldsymbol{\alpha}_r. \tag{4.42}$$

Once the values of the recovered gradient has been determined at the vertices and mid-side nodes, the recovery operator is uniquely defined as shown in Lemma 4.5 and the error estimator is defined as in Section 4.4.

4.6.3 Patch Recovery for Quadrilateral Elements

The extension of the patch recovery technique to quadrilateral elements requires only minor modifications from the procedure for triangles. The superconvergence results for quadrilateral elements of Lesaint and Zlamal [79] indicate that the gradient is superconvergent if sampled at the Gauss–Legendre quadrature points. As always, superconvergence holds only under stringent assumptions on the partition and the smoothness of the solution, which are generally not valid for adaptive refinement procedures. Nevertheless, the Gauss–Legendre points are selected for sampling the gradient. The cases $p = 1$ and $p = 2$ are illustrated in Figure 4.3(b) and (d).

The functions used in $\boldsymbol{p}(\boldsymbol{x})$ fitting the gradient are based on the space $\mathbb{Q}_p$ that was used to construct the finite element subspace X. The recovered gradient at vertex nodes is defined as in (4.39) while the value at the mid-side nodes is defined as in (4.42). The only remaining case is when the nodes are located on the interior of an element. The approach employed here consists of a simple averaging of the values obtained in fitting the gradient over each of the four patches associated with the vertex nodes of the element.

For affine elements, the conditions of Lemma 4.5 are easily verified for the procedure described above. However, the situation is less clear-cut for isoparametric elements, and in general the conditions will *not* be satisfied.

4.7 A CAUTIONARY TALE

Recovery-based estimators possess a number of attractive features that have led to their popularity. In particular, their ease of implementation, generality, and ability to produce quite accurate estimators have led to their widespread adoption, especially in the engineering community. However, the estimators also have drawbacks.

To illustrate the dangers in the indiscriminate use of such techniques, we present a simple example. The setting is chosen to be as simple as possible so that there is no question that the source of the problems is due to extraneous effects such as mesh distortions (the partition will be uniform), numerical quadrature (all integrals are evaluated exactly), nonsmooth coefficients (the data are smooth) or nonsmooth solution (the true solution is smooth). Furthermore, the example is even taken to be one-dimensional to avoid any question regarding mesh topology. Nevertheless, even in such an idealistic setting, it is possible to construct an example whereby the recovery-based estimators produce an estimated error of *zero*, while the actual error can be arbitrarily large.

It would be quite wrong to dismiss such a situation as being purely academic and having no relevance to practical computation of much more complicated phenomena in multidimensions. The point is that if such anomalies can be present in such a simplified setting, then it is quite possible for similar effects to occur, possibly only locally, in more complex cases. The danger would be that the estimated error in the neighborhood of the local feature would be zero, with the net result that the adaptive refinement procedure would miss the local feature altogether and the error estimator would not flag any difficulties. Obviously, this is a potentially disastrous situation.

The example consists of approximating the problem

$$-u'' = f \text{ on } I = (0,1); \quad u(0) = u(1) = 0$$

The data f is chosen to be of the form $f(x) = \mu \sin(2^m \pi x)$, where m is a fixed integer and μ is an arbitrary constant. The approximation consists of using piecewise linear finite elements on a uniform partition consisting of 2^n elements. Such a partition would arise beginning with a single element and performing $n-1$ successive subdivisions of the partition, corresponding to a sequence of uniform refinements. The element nodes are located at the points

$$x_k = k/2^n, \quad k = 0, \dots, 2^n.$$

It is well known that the finite element approximation u_h will actually be the piecewise linear interpolant to the exact solution u. Here, the exact solution is given by

$$u(x) = \frac{1}{4^m \pi^2} \mu \sin(2^m \pi x)$$

and if $m \geq n$, then $u(x_k)$ vanishes identically. Consequently, the finite element approximation u_h is identically zero. Observe that, no matter how many times the partition is subdivided by increasing n, there always exists a set of data (corresponding to sufficiently large m) such that u_h will vanish identically. Obviously, the gradient of the finite element approximation also vanishes everywhere, and consequently the recovered gradient $G_X(u_h)$ will also vanish everywhere. This means that the estimated error will be zero. However, the actual error is proportional to $|\mu|$ and could therefore be arbitrarily large.

4.8 BIBLIOGRAPHICAL REMARKS

The general framework for recovery operators and their application to *a posteriori* error estimation was given in Ainsworth and Craig [6]. An extensive survey of superconvergence results will be found in the article of Křížek and Neitaanmaki [76]. The patch recovery technique was originally proposed by Zienkiewicz and Zhu [122, 123].

In closing, we would like to emphasize that while the superconvergence property has been frequently used as a guideline in selecting the sampling points, the failure of the condition in situations of practical interest means that the analysis is somewhat unsatisfactory. However, it is worth pointing out that while the superconvergence property is generally lost, the accuracy and robustness of the associated recovery-based estimator remains highly satisfactory. At the present time, there is no fully satisfactory explanation behind the reason for this phenomenon. One possibility lies in the fact that it is by no means necessary for the gradient sampled at the superconvergence points to be converging at a higher rate. Instead, it is sufficient for the gradient sampled at such points to be more accurate by a multiplicative factor $\eta \ll 1$. This property has been referred to as $\eta\%$ superconvergence and investigated by Babuška *et al.* [26, 28]. It is found that this weaker condition does indeed continue to hold on less regular meshes of the type more likely to appear in practical computation, in certain circumstances. A second explanation for the surprising robustness of the estimators obtained from the patch recovery procedure has been investigated by Zhang and Zhu [119, 120], where the patch recovery procedure itself is shown to produce precisely those linear combinations of the sampled gradients that are needed to manufacture extra accuracy that would lead to more accurate error estimation.

5

Estimators, Indicators, and Hierarchic Bases

5.1 INTRODUCTION

One technique that is employed throughout numerical analysis to obtain computable error estimates consists of solving the problem of interest using two discretization schemes of differing accuracy and using the difference in the approximations as an estimate for the error. Examples of such schemes include the Runge–Kutta–Fehlberg method or step-length halving with Richardson extrapolation used in the numerical solution of ordinary differential equations [46, 70], Kronrod–Patterson or Romberg rules for automatic numerical quadrature [105], and deferred-correction techniques in general [104]. The attractiveness of such ideas stems from their applicability to quite general classes of problems combined with simplicity and ease of implementation. This chapter is concerned with their application to *a posteriori* error estimation in the context of finite element approximation.

Consider the elliptic boundary value problem of Section 1.4 and suppose the finite element approximation $u_X \in X$ is known. Let the finite-dimensional subspace $Y \subset V$ be an enrichment of the original finite element subspace X. Naturally, there is no advantage in duplicating functions from the original finite element subspace X in the enriched space Y, so it is assumed that these spaces satisfy

$$X \cap Y = \{0\}. \tag{5.1}$$

An improved approximation of the exact solution u is sought from the space $X^* = X \oplus Y$ by finding the Galerkin approximation

$$B(u^*, v^*) = L(v^*) \quad \forall v^* \in X^*. \tag{5.2}$$

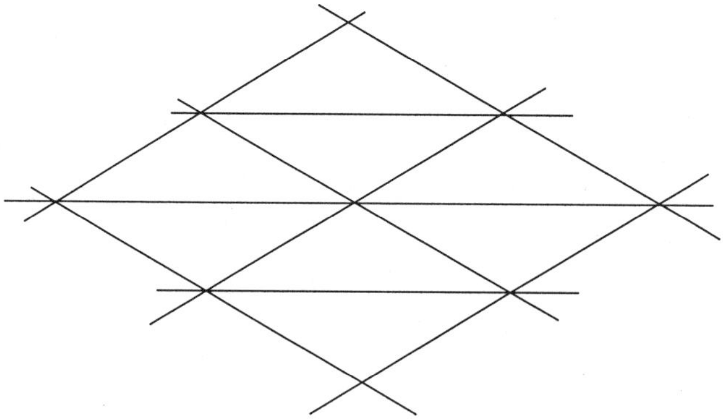

Fig. 5.1 Portion of original mesh used to construct piecewise linear finite element subspace X.

The enhanced space X^* may, for example, be constructed by augmenting the original space X with higher-order basis functions, or may simply consist of a uniform refinement of the mesh used to construct the space X. The theory will be illustrated for some common choices for the enhanced space Y when the underlying finite element subspace X consists of continuous piecewise linear functions on a mesh of triangular elements–for example, as shown in Figure 5.1.

The usual way to enhance the space X consists of either increasing the polynomial degree of the elements so that the enriched space X^* consists of piecewise quadratics, or, alternatively, refining the mesh uniformly so that X^* consists of piecewise linear functions on a refined mesh. A typical basis function for the space Y is shown in Figure 5.2 for the case when the enhanced space X^* consists of piecewise quadratic functions. The situation is less clear-cut if the enhanced space X^* is obtained by a uniform subdivision of the mesh, and two possible cases are shown in Figure 5.2.

If one is prepared to accept that the approximation u^* is superior to the original approximation, then

$$|||e||| = |||u - u_X||| \approx |||u^* - u_X||| = |||e^*||| \tag{5.3}$$

and the difference between the two approximations $u^* - u_X$ will provide a computable estimate for the error [57]. The precise meaning of this approximation will be the subject of the next section. The main drawback of this approach is the sheer expense of solving a problem of larger complexity than the problem leading to the finite element approximation itself. Therefore, methods are then explored for avoiding the full problem, and quantitative bounds for the resulting error estimators are derived.

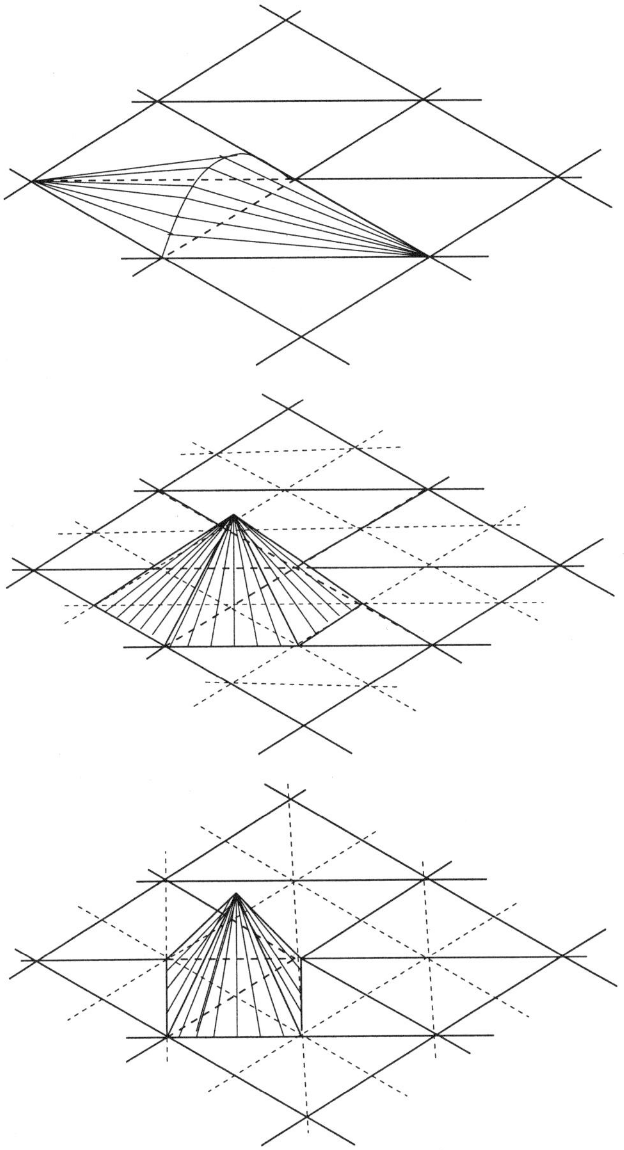

Fig. 5.2 Hierarchic basis functions for the space Y when the enhanced space X^* is constructed using piecewise quadratic functions, or piecewise linear functions on the meshes obtained using the regular subdivision (indicated in dashed lines) and the criss-cross subdivision.

5.2 SATURATION ASSUMPTION

The effectiveness of the approximation $|||e^*||| \approx |||e|||$ is dictated by whether u^* really does represent an improved approximation over the finite element approximation u_X. This notion is quantified in terms of the *saturation assumption* [36]: There exists a constant $\beta \in [0, 1)$ such that

$$|||u - u^*||| \leq \beta \, |||u - u_X||| \, . \tag{5.4}$$

The optimality of the Galerkin procedure means that

$$|||u - u^*||| = \inf_{v^* \in X^*} |||u - v^*||| \, ; \tag{5.5}$$

as a consequence, it follows that

$$|||u - u^*||| \leq |||u - u_X||| \, , \tag{5.6}$$

showing that condition (5.4) always holds for some $\beta \leq 1$. The real import of the saturation condition is that the constant β must be strictly less than unity uniformly.

It is easy to convince oneself that the saturation condition will hold for "reasonable" functions u. For example, suppose that the Galerkin subspace X consists of piecewise linear functions while the enhanced space is constructed by augmenting X with quadratic basis functions. If the solution u is smooth, say $u \in H^3(\Omega)$, then the standard *a priori* error estimates imply that on quasi-uniform meshes of size h, we have

$$|||u - u_X||| \leq Ch \, ||u||_{H^2(\Omega)} \tag{5.7}$$

and

$$|||u - u^*||| \leq Ch^2 \, ||u||_{H^3(\Omega)} \, . \tag{5.8}$$

Therefore, asymptotically, one would expect that $\beta = \mathcal{O}(h)$—a much stronger property than is required for the saturation assumption. Equally well, if the enhanced space X^* consists of a uniform subdivision of the mesh used to construct X, then the same *a priori* estimates lead one to suspect that for small mesh size h, $\beta \approx 1/2$.

Despite these arguments, the saturation assumption will fail to be true in general [42]. For instance, let $Z \subset L_2(\Omega)$ be any given subspace satisfying

$$\dim Z > \dim Y. \tag{5.9}$$

Define the linear mapping $\ell : Z \to X^*$ by the rule

$$\ell(g) = w^* - w_X \tag{5.10}$$

where w^* and w_X satisfy

$$w_X \in X : B(w_X, v_X) = (g, v_X) \quad \forall v_X \in X \tag{5.11}$$

and

$$w^* \in X^* : B(w^*, v^*) = (g, v^*) \quad \forall v^* \in X^*. \tag{5.12}$$

Thus, $\ell(g) \in X^\perp \cap X^*$, where $X^\perp$ is the orthogonal complement of X with respect to the energy inner product, and so

$$\dim Z > \dim Y = \dim \left(X^\perp \cap X^* \right). \tag{5.13}$$

Therefore, $\ell : Z \to X^\perp \cap X^*$ is not one to one and there must exist a nontrivial $f \in Z$ such that $\ell(f)$ vanishes. That is to say,

$$0 = \ell(f) = u^* - u_X \tag{5.14}$$

and so u^* and u_X coincide. Hence, unless $u \in X$, the constant in the saturation assumption satisfies $\beta = 1$.

This example shows that the saturation assumption is invalid for large classes of data f. This failure is to be expected since u^* and u_X belong to finite-dimensional spaces and so the component of the data f that is orthogonal (in an L_2 sense) to the spaces is essentially "invisible". Fortunately, the data in practical computations are taken from quite restricted sets such as piecewise polynomial or piecewise analytic functions, so that the saturation assumption may be quite realistic in a practical setting.

5.3 ANALYSIS OF ESTIMATOR

Our point of departure will be the residual equation satisfied by the true error

$$e \in V : \quad B(e, v) = L(v) - B(u, v) \quad v \in V. \tag{5.15}$$

Let $X^* = X \oplus Y$ be an enhanced subspace as above. One possibility that immediately suggests itself is to approximate the error directly by constructing its Galerkin approximation e^* from the enhanced space X^*,

$$B(e^*, v^*) = L(v^*) - B(u_X, v^*) \quad \forall v^* \in X^*. \tag{5.16}$$

It is elementary to verify that e^* is related to the enhanced approximation u^* by the equation $e^* = u^* - u_X$. However, reformulating the problem in this way will be convenient for subsequent developments.

The first result shows how the accuracy of the error estimator is related to the saturation assumption.

Theorem 5.1 *Suppose that the saturation condition (5.4) holds for the solution u. Then, the estimator defined by (5.16) satisfies*

$$|||e^*||| \leq |||e||| \leq \frac{1}{\sqrt{1-\beta^2}} |||e^*|||. \tag{5.17}$$

Proof. Observe that

$$e = u - u_X = (u - u^*) + e^*$$

and

$$B(u - u^*, e^*) = L(e^*) - L(e^*) = 0,$$

so the Pythagorean identity gives

$$|||e|||^2 = |||u - u^*|||^2 + |||e^*|||^2 .$$

The left-hand inequality follows at once since $|||u - u^*||| \geq 0$. The right-hand inequality is obtained by invoking the saturation assumption and then rearranging. ■

5.4 ERROR ESTIMATION USING A REDUCED SUBSPACE

The main drawback of the estimator, assuming that the saturation assumption is valid, lies in the expense entailed in computing e^* from the larger space X^*. The problem may be regarded as consisting of finding the components of e^* from the spaces X and Y:

$$e^* = e_X^* + e_Y^* \tag{5.18}$$

where $e_X^* \in X$ and $e_Y^* \in Y$ satisfy

$$\begin{aligned} B(e_X^*, v_X) + B(e_Y^*, v_X) &= L(v_X) - B(u_X, v_X) = 0 \quad \forall v_X \in X \\ B(e_X^*, v_Y) + B(e_Y^*, v_Y) &= L(v_Y) - B(u_X, v_Y) \quad \forall v_Y \in Y \end{aligned}$$

and the right-hand side in the first equation vanishes, thanks to u_X being the Galerkin approximation from the original space X. In search of a reasonable approximation $\overline{e}$ to e^*, one may ignore the coupling terms between the spaces in the system, and seek $\overline{e} = \overline{e}_X + \overline{e}_Y$ such that

$$\begin{aligned} B(\overline{e}_X, v_X) &= 0 \quad \forall v_X \in X \\ B(\overline{e}_Y, v_Y) &= L(v_Y) - B(u_X, v_Y) \quad \forall v_Y \in Y. \end{aligned}$$

Evidently, $\overline{e}_X$ vanishes and one is therefore only obliged to solve a problem for $\overline{e}_Y$, or equally well, $\overline{e}$, on the subspace Y.

In summary, a modified error estimator is obtained by solving the residual problem over the reduced space Y,

$$B(\overline{e}, v) = L(v) - B(u_X, v) \quad \forall v \in Y \tag{5.19}$$

and the estimator is taken to be

$$|||e||| \approx |||\overline{e}||| . \tag{5.20}$$

For the purposes of analysis, it is useful to note that these approximations to the true error are related by the equations

$$\begin{aligned} B(e_X^*, v_X) + B(e_Y^*, v_X) &= 0 \quad \forall v_X \in X \\ B(e_X^*, v_Y) + B(e_Y^*, v_Y) &= B(\bar{e}, v_Y) \quad \forall v_Y \in Y. \end{aligned} \tag{5.21}$$

The accuracy of the estimator will depend on the loss incurred by neglecting the coupling terms in the full problem. It is illuminating to first consider the simple case when X and Y are both one-dimensional:

$$X = \text{span}\{\phi_X\}; \quad Y = \text{span}\{\phi_Y\}. \tag{5.22}$$

The system reduces to a matrix equation

$$\begin{bmatrix} B(\phi_X, \phi_X) & B(\phi_X, \phi_Y) \\ B(\phi_Y, \phi_X) & B(\phi_Y, \phi_Y) \end{bmatrix} \begin{bmatrix} e_X^* \\ e_Y^* \end{bmatrix} = \begin{bmatrix} 0 \\ B(\bar{e}, \phi_Y) \end{bmatrix} \tag{5.23}$$

and it is readily found that the energies of the solutions are related by

$$|||e^*|||^2 = \frac{1}{1-\gamma^2} |||\bar{e}|||^2 \tag{5.24}$$

where

$$\gamma = \frac{|B(\phi_X, \phi_Y)|}{|||\phi_X|||\;|||\phi_Y|||}. \tag{5.25}$$

The parameter γ measures the (cosine of the) angle between the subspaces X and Y. For instance, the Cauchy–Schwarz inequality immediately shows that $\gamma \leq 1$, with equality holding if and only if ϕ_X and ϕ_Y are linearly dependent. However, since the spaces X and Y share only the zero element, this means that $\gamma < 1$. If the subspaces are orthogonal, then $\gamma = 0$ and $|||e^*||| = |||\bar{e}|||$ as would be expected, since there is no coupling between the subspaces in the first instance.

This simple example indicates that some notion of orthogonality will be needed in the general case. The assumption made is that a *strengthened Cauchy–Schwarz inequality* holds [36]; that is, there exists a constant γ such that

$$\sup\{|B(v_X, v_Y)| : v_X \in X : |||v_X||| = 1, v_Y \in Y : |||v_Y||| = 1\} \leq \gamma < 1 \tag{5.26}$$

or, equally well,

$$|B(v_X, v_Y)| \leq \gamma\, |||v_X|||\; |||v_Y||| \quad \forall v_X \in X, v_Y \in Y. \tag{5.27}$$

The following result concerning the accuracy of the error estimator in the general case mirrors the simple example discussed above. Moreover, the simple example shows that the result cannot be improved.

Theorem 5.2 *Suppose that there exists a constant $\gamma \in [0,1)$ such that the strengthened Cauchy Schwarz inequality (5.26) holds. Then, the error estimator defined in equations (5.19)–(5.20) satisfies*

$$|||\bar{e}||| \leq |||e^*||| \leq \frac{1}{\sqrt{1-\gamma^2}} |||\bar{e}||| . \tag{5.28}$$

Proof. Applying the strengthened Cauchy–Schwarz inequality,

$$|||e^*|||^2 = |||e^*_X|||^2 + 2B(e^*_X, e^*_Y) + |||e^*_Y|||^2 \geq |||e^*_X|||^2 - 2\gamma |||e^*_X||| \, |||e^*_Y||| + |||e^*_Y|||^2$$

and then using the inequality

$$2\gamma |||e^*_X||| \, |||e^*_Y||| \leq |||e^*_X|||^2 + \gamma^2 |||e^*_Y|||^2$$

gives

$$|||e^*|||^2 \geq (1-\gamma^2) |||e^*_Y|||^2 .$$

Now, applying the first equation in the system (5.21) yields

$$|||e^*|||^2 = B(e^*, e^*_X) + B(e^*, e^*_Y) = B(e^*, e^*_Y)$$

and then applying the second equation in (5.21) gives

$$|||e^*|||^2 = B(\bar{e}, e^*_Y) \leq |||\bar{e}||| \, |||e^*_Y||| .$$

The right-hand inequality now follows since

$$|||e^*|||^2 \leq |||\bar{e}||| \, |||e^*_Y||| \leq \frac{1}{\sqrt{1-\gamma^2}} |||\bar{e}||| \, |||e^*||| .$$

The left-hand inequality is obtained from the second equation in (5.21) since

$$|||\bar{e}|||^2 = B(\bar{e}, \bar{e}) = B(e^*, \bar{e}) \leq |||\bar{e}||| \, |||e^*|||$$

and the proof is complete. ■

The estimator defined by (5.19) requires the inversion of a global stiffness matrix for the space Y—at a cost that may well prove to be prohibitively expensive.

This naturally leads us to consider the use of an inexact solver in lieu of the exact solution of the linear system. A convenient inexact solver might consist of a single sweep of an iterative solution method. In particular, a single sweep of the Jacobi iteration corresponds to approximating the full matrix by its diagonal. Alternatively, a single multigrid cycle could be used. A convenient theoretical framework for the analysis of inexact solvers may be developed by interpreting the approximate solution of the matrix equation involving the stiffness matrix for bilinear form $B(\cdot,\cdot)$ and being the exact solution of a matrix equation arising from an approximate bilinear form. Thus, the inexact solver is equivalent to inserting an alternative inner product $A(\cdot,\cdot)$ that is more easily inverted, in place of the true bilinear form $B(\cdot,\cdot)$. The error estimator is obtained by solving the problem (5.19) with the new inner product in place of the original bilinear form:

$$\tilde{e} \in Y : A(\tilde{e}, v) = L(v) - B(u_X, v) \quad \forall v \in Y. \tag{5.29}$$

The error estimator could then be taken to be

$$|||e||| \approx |||\tilde{e}||| . \tag{5.30}$$

A more attractive practical alternative would consists of taking the estimator to be

$$|||e||| \approx |||\tilde{e}|||_A \tag{5.31}$$

where $|||\cdot|||_A$ is the approximate energy norm based on the approximate bilinear form $A(\cdot,\cdot)$.

The accuracy of the resulting error estimator, compared with the estimator using the true inner product, depends on how well the new inner product respects the spectrum of the original one.

Theorem 5.3 *Suppose that there exist positive quantities λ and Λ such that*

$$\lambda B(v,v) \leq A(v,v) \leq \Lambda B(v,v) \quad \forall v \in Y. \tag{5.32}$$

Then, the error estimator defined in equations (5.29)–(5.30) satisfies

$$\lambda |||\tilde{e}||| \leq |||\overline{e}||| \leq \Lambda |||\tilde{e}||| \tag{5.33}$$

while the estimator defined by equations (5.31) satisfies

$$\sqrt{\lambda}\, |||\tilde{e}|||_A \leq |||\overline{e}||| \leq \sqrt{\Lambda}\, |||\tilde{e}|||_A . \tag{5.34}$$

Proof. The approximations to the true error are related by the condition

$$A(\tilde{e}, v) = B(\overline{e}, v) \quad \forall v \in Y.$$

First, select $v = \overline{e}$ and apply the Cauchy–Schwarz inequality to obtain

$$|||\overline{e}|||^2 = B(\overline{e},\overline{e}) = A(\tilde{e},\overline{e}) \leq A(\tilde{e},\tilde{e})^{1/2} A(\overline{e},\overline{e})^{1/2} \leq \Lambda\, |||\overline{e}|||\, |||\tilde{e}|||$$

and the left-hand inequality follows. Now, choosing $v = \tilde{e}$ gives

$$\lambda |||\tilde{e}|||^2 \leq A(\tilde{e},\tilde{e}) = B(\overline{e},\tilde{e}) \leq |||\overline{e}|||\, |||\tilde{e}|||$$

and the right-hand inequality follows. ■

It is interesting to observe that the estimator obtained by evaluating the true energy of the approximate solution is inferior to the estimator arising from evaluating the energy using the approximate inner product. Consequently, the approximate inner product is to be preferred from both the theoretical and practical standpoint.

5.5 THE STRENGTHENED CAUCHY–SCHWARZ INEQUALITY

The general validity of the strengthened Cauchy–Schwarz inequality is confirmed by the following result [63].

Theorem 5.4 *Let V be a Hilbert space equipped with the inner product $(\cdot,\cdot)_V$ and norm denoted by $\|\cdot\|_V$ and let X, Y be a pair of finite-dimensional subspaces satisfying $X \cap Y = \{0\}$. Then, there exists a constant $\gamma \in [0,1)$, depending on the spaces X and Y, such that for all $x \in X$ and $y \in Y$,*

$$(x,y)_V \leq \gamma \|x\|_V \|y\|_V \tag{5.35}$$

where $\|\cdot\|_V$ denotes the norm associated with the inner product $(\cdot,\cdot)_V$.

Proof. Let

$$\gamma = \sup_{x\in X, y\in Y} \frac{(x,y)_V}{\|x\|_V \|y\|_V};$$

then $\gamma \leq 1$ by the usual Cauchy–Schwarz inequality.

Suppose that the result claimed is false, and $\gamma = 1$. Then there would exist sequences $\{x_n\} \in X$ and $\{y_n\} \in Y$ satisfying

- $\|x_n\|_V = \|y_n\|_V = 1$ for all n
- $(x_n, y_n)_V \to 1$ as $n \to \infty$.

The compactness of the unit ball in finite-dimensional spaces means that each of the sequences has a convergent subsequence with limits $x \in X$ and $y \in Y$ satisfying $\|x\|_V = \|y\|_V = 1$. The continuity of the inner product implies that $(x,y)_V = 1$. Consequently,

$$(x,y)_V = 1 = \|x\|_V \|y\|_V$$

so that the Cauchy–Schwarz inequality holds as an equality. However, it is known that this is possible if and only if x and y are linearly dependent. Therefore, since $X \cap Y = \{0\}$, it follows that $x = y = 0$. However, this is a contradiction of the fact $\|x\|_V = \|y\|_V = 1$. ■

The strengthened Cauchy–Schwarz inequality is equivalent to the boundedness of a certain projection operator. Specifically, each $u \in X^*$ may be decomposed into contributions from the spaces X and Y,

$$u = u_X + u_Y. \tag{5.36}$$

The uniqueness of this decomposition ensures that the operator $\Pi : X^* \to X$ given by the rule

$$\Pi u = u_X \tag{5.37}$$

is well-defined. In particular, if $u_X \in X$, then $\Pi u_X = u_X$ and

$$\Pi \circ \Pi u = \Pi u_X = u_X = \Pi u, \tag{5.38}$$

thereby showing Π is a projection. Similarly, if $u_Y \in Y$ then $\Pi u_Y = 0$.

The following result was given in [107]:

Theorem 5.5 *Suppose $X^* = X + Y$ where $X \cap Y = \{0\}$, and let $\Pi : X^* \to X$ be defined as above. Then the following statements are equivalent:*

1. *There exists a constant $M < \infty$ such that*

$$\|\Pi u\| \leq M \|u\| \quad \forall u \in X^*. \tag{5.39}$$

2. *There exists a constant $\gamma \in [0, 1)$ such that*

$$B(x, y) \leq \gamma \|x\| \|y\| \quad \forall x \in X, y \in Y. \tag{5.40}$$

Moreover, the constants are related by the equation

$$\gamma^2 = 1 - \frac{1}{M^2}. \tag{5.41}$$

Proof. (1) $\Rightarrow$ (2): Let $x \in X$ and $y \in Y$ be given, and for $t \in \mathbb{R}$, define $x_t \in X^*$ to be

$$x_t = x + ty. \tag{5.42}$$

From the earlier remarks, $\Pi x = x$ and $\Pi y = 0$ and hence $\Pi x_t = x$. Therefore,

$$\|x\|^2 = \|\Pi x_t\|^2 \leq M^2 \|x_t\|^2$$

or, on expanding the right-hand side and simplifying,

$$t^2 \|y\|^2 + 2tB(x, y) + \left(1 - \frac{1}{M^2}\right) \|x\|^2 \geq 0.$$

The nonnegativity of this quadratic form in t means that the discriminant must be positive, or equally well,

$$B(x, y)^2 \leq \left(1 - \frac{1}{M^2}\right) \|x\|^2 \|y\|^2$$

and it follows that the strengthened Cauchy–Schwarz inequality holds with constant $\gamma^2 = 1 - 1/M^2$.

(2) $\Rightarrow$ (1): Let $u \in X^*$ be decomposed in the form $u = \Pi u + (I - \Pi)u$. Then,

$$\begin{aligned} \|u\|^2 &= \|\Pi u\|^2 + 2B(\Pi u, (I - \Pi)u) + \|(I - \Pi)u\|^2 \\ &\geq \|\Pi u\|^2 - 2\gamma \|\Pi u\| \|(I - \Pi)u\| + \|(I - \Pi)u\|^2 \\ &\geq (1 - \gamma^2) \|\Pi u\|^2 \end{aligned}$$

where the inequality

$$2\gamma \|\Pi u\| \, \|(I-\Pi)u\| \leq \gamma^2 \|\Pi u\|^2 + \|(I-\Pi)u\|^2$$

has been used. Hence, since $\gamma \in [0,1)$, the operator Π is bounded with norm at most $(1-\gamma^2)^{-1/2}$. ∎

The first of the above results shows that the strengthened Cauchy–Schwarz inequality holds under quite general conditions but provides no quantitative information on the size of the constant γ. The previous result shows how the magnitude of the constant is related to the norm of the projection operator Π. This fact will now be exploited in developing a technique for computing the value of the constant [83].

Let X and Y be finite-dimensional subspaces and introduce bases for the spaces

$$X = \text{span}\{\phi_k : k \in \mathcal{N}_X\} \tag{5.43}$$

and

$$Y = \text{span}\{\phi_k : k \in \mathcal{N}_Y\} \tag{5.44}$$

where $\mathcal{N}_X$ and $\mathcal{N}_Y$ are suitable indexing sets for the spaces X and Y, respectively. Any function $u \in X^* = X \oplus Y$ may be expanded in terms of the bases so that

$$u = \sum_{k \in \mathcal{N}_X \cup \mathcal{N}_Y} u_k \phi_k \tag{5.45}$$

with

$$\Pi u = u_X = \sum_{k \in \mathcal{N}_X} u_k \phi_k \tag{5.46}$$

and

$$u - \Pi u = u_Y = \sum_{k \in \mathcal{N}_Y} u_k \phi_k . \tag{5.47}$$

The energy of the function u may be written in terms of the stiffness matrix $\boldsymbol{B}$ for the enhanced approximation from the full space X^*,

$$\|u\|^2 = \boldsymbol{u}^\top \boldsymbol{B} \boldsymbol{u} \tag{5.48}$$

where $\boldsymbol{u}$ is the vector composed of the coefficients in the expansion (5.45). Alternatively, if the vector $\boldsymbol{u}$ is partitioned according to the coefficients in the expansions (5.46) and (5.47), then

$$\|u\|^2 = \begin{bmatrix} \boldsymbol{u}_X \\ \boldsymbol{u}_Y \end{bmatrix}^\top \boldsymbol{B} \begin{bmatrix} \boldsymbol{u}_X \\ \boldsymbol{u}_Y \end{bmatrix} \tag{5.49}$$

and, similarly, the energy of the projection Πu is easily seen to be

$$\|\Pi u\|^2 = \begin{bmatrix} \boldsymbol{u}_X \\ \boldsymbol{0} \end{bmatrix}^\top \boldsymbol{B} \begin{bmatrix} \boldsymbol{u}_X \\ \boldsymbol{0} \end{bmatrix} . \tag{5.50}$$

The stiffness matrix may be partitioned into blocks respecting the decomposition $X^* = X \oplus Y$,

$$\boldsymbol{B} = \begin{bmatrix} \boldsymbol{B}_{XX} & \boldsymbol{B}_{XY} \\ \boldsymbol{B}_{YX} & \boldsymbol{B}_{YY} \end{bmatrix}. \tag{5.51}$$

The main result concerning the numerical evaluation of the constant in the strengthened Cauchy–Schwarz inequality may now be stated.

Theorem 5.6 *With the above notation and assumptions, the constant in the strengthened Cauchy–Schwarz inequality is given by*

$$\gamma^2 = \max_{\boldsymbol{x}} \frac{\boldsymbol{x}^\top \boldsymbol{B}_{XY} \boldsymbol{B}_{YY}^{-1} \boldsymbol{B}_{YX} \boldsymbol{x}}{\boldsymbol{x}^\top \boldsymbol{B}_{XX} \boldsymbol{x}}. \tag{5.52}$$

Proof. The norm of the operator Π is defined to be

$$M = \sup_{u \in X^*} \frac{|||\Pi u|||}{|||u|||}$$

and therefore, the foregoing arguments show that

$$M^2 = \sup_{\boldsymbol{u}_X} \sup_{\boldsymbol{u}_Y} \frac{\boldsymbol{u}_X^\top \boldsymbol{B}_{XX} \boldsymbol{u}_X}{\boldsymbol{u}^\top \boldsymbol{B} \boldsymbol{u}}. \tag{5.53}$$

Elementary manipulations reveal that

$$\boldsymbol{u}^\top \boldsymbol{B} \boldsymbol{u} = \boldsymbol{u}_X^\top \boldsymbol{S}_{XX} \boldsymbol{u}_X + (\boldsymbol{u}_Y + \boldsymbol{B}_{YY}^{-1} \boldsymbol{B}_{YX} \boldsymbol{u}_X)^\top \boldsymbol{B}_{YY} (\boldsymbol{u}_Y + \boldsymbol{B}_{YY}^{-1} \boldsymbol{B}_{YX} \boldsymbol{u}_X)$$

where

$$\boldsymbol{S}_{XX} = \boldsymbol{B}_{XX} - \boldsymbol{B}_{XY} \boldsymbol{B}_{YY}^{-1} \boldsymbol{B}_{YX}.$$

Therefore, the supremum over $\boldsymbol{u}_Y$ in (5.53) is attained when the second term vanishes and $\boldsymbol{u}_Y$ satisfies

$$\boldsymbol{u}_Y + \boldsymbol{B}_{YY}^{-1} \boldsymbol{B}_{YX} \boldsymbol{u}_X = \boldsymbol{0}.$$

Hence,

$$M^2 = \sup_{\boldsymbol{u}_X} \frac{\boldsymbol{u}_X^\top \boldsymbol{B}_{XX} \boldsymbol{u}_X}{\boldsymbol{u}_X^\top \boldsymbol{S}_{XX} \boldsymbol{u}_X}$$

and, after replacing $\boldsymbol{S}_{XX}$ and simplifying, one easily arrives at

$$1 - \frac{1}{M^2} = \sup_{\boldsymbol{u}_X} \frac{\boldsymbol{u}_X^\top \boldsymbol{B}_{XY} \boldsymbol{B}_{YY}^{-1} \boldsymbol{B}_{YX} \boldsymbol{u}_X}{\boldsymbol{u}_X^\top \boldsymbol{B}_{XX} \boldsymbol{u}_X}$$

and the result follows as claimed on recalling Theorem 5.1. ■

5.6 EXAMPLES

The foregoing theory has shown that the estimators satisfy

$$\lambda \|\tilde{e}\|_A^2 \leq \|\bar{e}\|^2 \leq \|e^*\|^2 \leq \|e\|^2 \tag{5.54}$$

and

$$\|e\|^2 \leq \frac{\|e^*\|^2}{1-\beta^2} \leq \frac{1}{1-\gamma^2}\frac{\|\bar{e}\|^2}{1-\beta^2} \leq \frac{\Lambda}{1-\gamma^2}\frac{\|\tilde{e}\|_A^2}{1-\beta^2}. \tag{5.55}$$

The performance of the estimators depends on the values of the parameters appearing in these relations. Earlier remarks indicate that the value of β depends on the true solution u of the particular problem, and one has little control over its value in general. However, the parameters Λ, λ and γ depend only on the bilinear form and the spaces X and Y, and may be computed explicitly, thereby enabling quantitative analysis of the performance of the estimators.

A useful observation is that the constant in the Cauchy–Schwarz inequality may be deduced using computations on a single element. For instance, suppose that the constant γ_K is known for each element K in the mesh,

$$B_K(u,v) \leq \gamma_K \|u\|_K \|v\|_K \quad \forall u \in X_K, v \in Y_K \tag{5.56}$$

where the subscript indicates restriction to the element K. The constant γ for the whole mesh is obtained by summation

$$\begin{aligned} B(u,v) &= \sum_K B_K(u,v) \\ &\leq \sum_K \gamma_K \|u\|_K \|v\|_K \\ &\leq \gamma \sum_K \|u\|_K \|v\|_K \\ &\leq \gamma \|u\| \|v\| \end{aligned}$$

where $\gamma = \max \gamma_K$, and the discrete Cauchy–Schwarz inequality has been employed.

The case when the enhanced space X^* is constructed using quadratic functions will now be considered in some detail for the case of approximation of Poisson's equation. The basis functions for the original space X on the triangle with vertices at $(0,0)$, $(1,0)$, and $(0,1)$ are given by

$$\begin{aligned} \phi_1(x,y) &= 1-x-y \\ \phi_2(x,y) &= x \\ \phi_3(x,y) &= y \end{aligned}$$

while a basis for the space Y consists of the quadratic bubble functions

$$\begin{aligned}\phi_4(x,y) &= (1-x-y)x\\ \phi_5(x,y) &= xy\\ \phi_6(x,y) &= (1-x-y)y.\end{aligned}$$

The stiffness matrix for the bilinear form $B(\cdot,\cdot)$ relative to these basis functions is

$$\boldsymbol{B} = \begin{bmatrix} 1 & -1/2 & -1/2 & -1/3 & 1/6 & 1/6 \\ -1/2 & 1/2 & 0 & 1/6 & -1/6 & 0 \\ -1/2 & 0 & 1/2 & 1/6 & 0 & -1/6 \\ -1/3 & 1/6 & 1/6 & 1/6 & -1/12 & -1/12 \\ 1/6 & -1/6 & 0 & -1/12 & 1/6 & 0 \\ 1/6 & 0 & -1/6 & -1/12 & 0 & 1/6 \end{bmatrix} \tag{5.57}$$

from where relevant subblocks are easily identified, and hence

$$\boldsymbol{B}_{XY}\boldsymbol{B}_{YY}^{-1}\boldsymbol{B}_{YX} = \begin{bmatrix} 2/3 & -1/3 & -1/3 \\ -1/3 & 1/4 & 1/12 \\ -1/3 & 1/12 & 1/4 \end{bmatrix}. \tag{5.58}$$

In view of Theorem 5.6, the strengthened Cauchy–Schwarz constant γ^2 is the maximum eigenvalue of the problem

$$\boldsymbol{B}_{XY}\boldsymbol{B}_{YY}^{-1}\boldsymbol{B}_{YX}\boldsymbol{x} = \mu \boldsymbol{B}_{XX}\boldsymbol{x}; \tag{5.59}$$

therefore, solving the problem leads to the conclusion that $\gamma^2 = 2/3$, in agreement with references [44] and [83]. A similar computation using the two alternative subspaces constructed using the regular and criss-cross subdivisions indicated in Figure 5.2 results in $\gamma^2 = 1/2$ in both cases, and, for the latter case, this is again consistent with references [44] and [83].

The constants λ and Λ in the hypothesis (5.32) are also easily obtained with an inexact solver, consisting of the diagonal $\boldsymbol{A}_{YY}$ of the matrix $\boldsymbol{B}_{YY}$, is used by computing the eigenvalues of the problem

$$\boldsymbol{B}_{YY}\boldsymbol{y} = \mu \boldsymbol{A}_{YY}\boldsymbol{y}. \tag{5.60}$$

If the quadratic space is used, then the extreme eigenvalues of this problem are found to be $1 \pm 1/\sqrt{2}$, so

$$(1-1/\sqrt{2})\boldsymbol{y}^\top \boldsymbol{A}_{YY}\boldsymbol{y} \le \boldsymbol{y}^\top \boldsymbol{B}_{YY}\boldsymbol{y} \le (1+1/\sqrt{2})\boldsymbol{y}^\top \boldsymbol{A}_{YY}\boldsymbol{y} \quad \forall \boldsymbol{y}. \tag{5.61}$$

or, in terms of bilinear forms,

$$(1-1/\sqrt{2})A_K(v,v) \le B_K(v,v) \le (1+1/\sqrt{2})A_K(v,v) \quad \forall v \in Y_K. \tag{5.62}$$

Summing these inequalities over all the elements reveals that hypothesis (5.32) is valid with the values $\lambda = (1-1/\sqrt{2})^{-1}$ and $\Lambda = (1+1/\sqrt{2})^{-1}$. As a matter of fact, the corresponding constants for the two alternative subdivision schemes are also found to be $(1 \pm 1/\sqrt{2})^{-1}$.

These results would suggest that there is little to choose between the various possible schemes. However, a different picture emerges if the bilinear form is generalized to

$$B(u,v) = \int_{\Omega} \nabla u^{\top} \begin{bmatrix} t\cos^2\theta + \sin^2\theta & (t-1)\cos\theta\sin\theta \\ (t-1)\cos\theta\sin\theta & \cos^2\theta + t\sin^2\theta \end{bmatrix} \nabla v \, dx \tag{5.63}$$

where the angle θ corresponds to the orientation of the mesh relative to the principal axes along which the permeability takes the values t and 1. The computation of the strengthened Cauchy–Schwarz constant γ^2 and the quantities λ and Λ is now more complicated. However, the use of symbolic manipulation packages allows such computations to be easily automated.

The variation of the Cauchy–Schwarz constant for each of the three enhanced spaces as the values of t and θ are varied is shown in Figures 5.3–5.5. The figures indicate that the constant γ^2 degenerates to unity severely as $t \to 0$ for certain values of the angle θ for both the quadratic space and the piecewise linear hierarchic space on the criss-cross subdivision. In contrast, the value of the constant for the piecewise linear hierarchic space on the regular subdivision is 3/4 of the value for the quadratic space [35], and in particular, γ^2 remains *uniformly bounded away from unity*. Thus, from the point of view of robustness, the piecewise linear hierarchic space on a regular subdivision is to be recommended.

5.7 MULTILEVEL ERROR INDICATORS

The principal drawback of the error estimators discussed so far lies in the limited approximation properties of the enhanced space X^*, manifested by the need to introduce the saturation assumption. One might ask whether it is possible to remedy this problem by further enhancements of the space X. For instance, in addition to augmenting the space X by the space Y_1 consisting of the hierarchical functions obtained through a single uniform refinement of the original mesh, further refinements may be employed to generate a sequence of enhancements $Y_1, \ldots, Y_\ell$, leading to

$$X_\ell^* = X \oplus Y_1 \oplus Y_2 \oplus \cdots \oplus Y_\ell. \tag{5.64}$$

The associated error estimator is obtained by solving

$$e^* \in X_\ell^* : B(e^*, v^*) = (f, v^*) - B(u_X, v^*) \quad \forall v^* \in X_\ell^* \tag{5.65}$$

and then, assuming the saturation condition

$$|||u - u^*||| \le \beta_\ell \, |||u - u_X||| \tag{5.66}$$

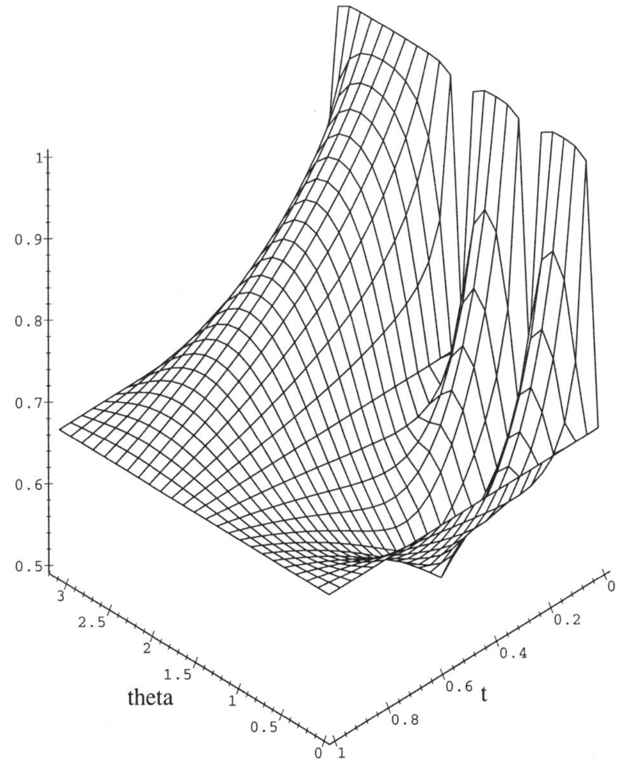

Fig. 5.3 Variation of the constant γ^2 in the strengthened Cauchy–Schwarz inequality for the bilinear form defined in (5.63) for the quadratic space.

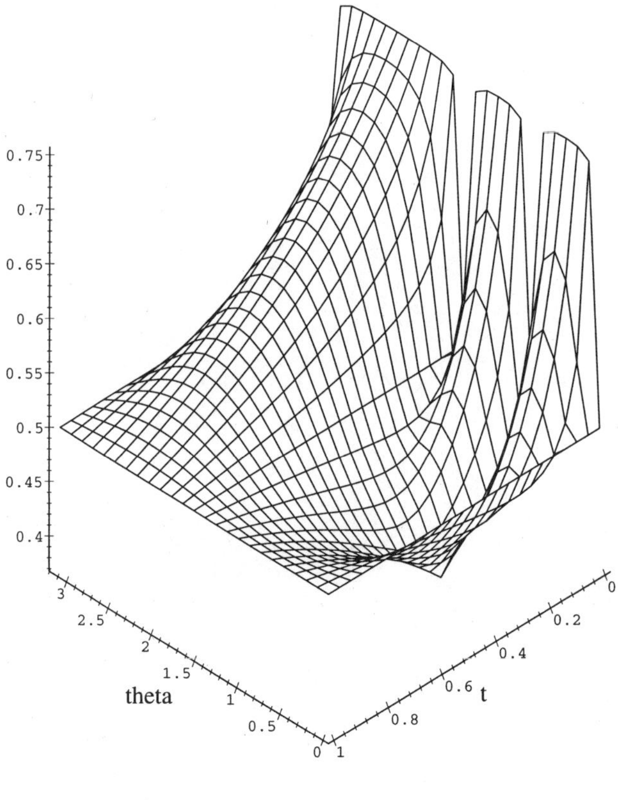

Fig. 5.4 Variation of the constant γ^2 in the strengthened Cauchy–Schwarz inequality for the bilinear form defined in (5.63) for piecewise linear hierarchic space on the regular subdivision.

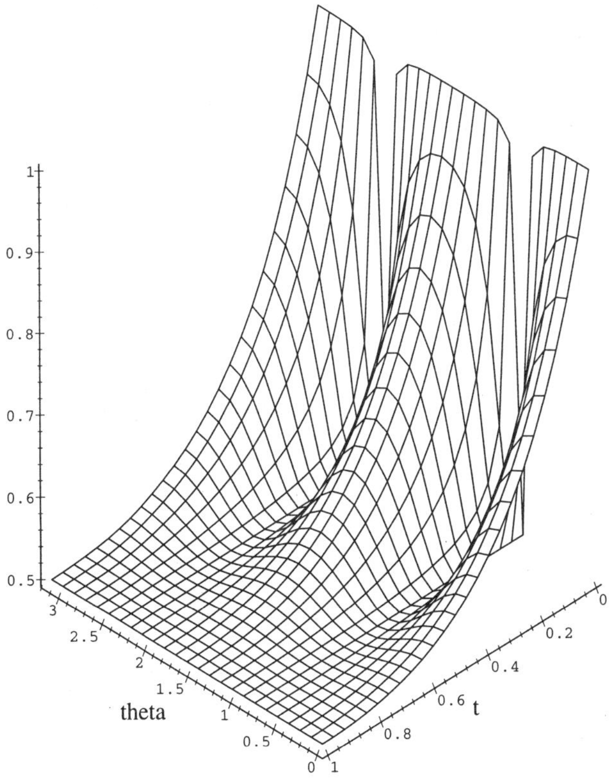

Fig. 5.5 Variation of the constant γ^2 in the strengthened Cauchy–Schwarz inequality for the bilinear form defined in (5.63) for piecewise linear hierarchic space on the criss-cross subdivision.

holds for some constant $\beta_\ell < 1$, gives the following result as in Theorem 5.1:

$$|||e^*||| \leq |||e||| \leq \frac{1}{\sqrt{1-\beta_\ell^2}} |||e^*||| . \tag{5.67}$$

The sequence of subspaces consisting of continuous piecewise linear functions on successive uniform subdivisions of the original mesh is dense in $H^1(\Omega)$, and consequently, $\beta_\ell \to 0$ in the limit $\ell \to \infty$. Obviously, it is not possible to take ℓ to infinity in practice. Nevertheless, the conclusion to be drawn is that the saturation condition may be satisfied by taking ℓ sufficiently large. The issue of what value of ℓ constitutes *sufficiently large* will be considered later.

As before, the approximation properties of this space will not be improved by duplication of functions, and it will therefore be assumed that

$$X \cap Y_k = \{0\}; \quad Y_j \cap Y_k = \{0\} \quad j \neq k. \tag{5.68}$$

The basis functions for the space Y_1 consist of the nodal basis functions at the nodes generated by a single refinement of the original mesh. By analogy, the basis functions for the space Y_k consist of the nodal basis functions at the nodes generated by the kth refinement of the original mesh. Figure 5.6 illustrates the basis functions in each of the spaces $\{Y_k : k \in \mathbb{N}\}$ for a one-dimensional domain.

At this stage, the selection of the particular basis has no influence on the error estimator $|||e^*|||$ since the computation takes place over the full space. However, the solution of the problem over the full space is enormously expensive and, as before, we shall seek to reduce the size of the computation. The problem (5.65) may be reformulated as a coupled system involving the contributions from each of the spaces by first writing

$$e^* = e_X^* + e_1^* + \cdots + e_\ell^* \tag{5.69}$$

where $e_X^* \in X$ and, $e_k^* \in Y_k$, $k = 1, \ldots, \ell$, satisfy

$$\left.\begin{aligned} B(e_X^*, v_X) + B(e_1^*, v_X) + \cdots + B(e_\ell^*, v_X) &= 0 \\ B(e_X^*, v_1) + B(e_1^*, v_1) + \cdots + B(e_\ell^*, v_1) &= (f, v_1) - B(u_X, v_1) \\ &\vdots \\ B(e_X^*, v_\ell) + B(e_\ell^*, v_\ell) + \cdots + B(e_\ell^*, v_\ell) &= (f, v_\ell) - B(u_X, v_\ell) \end{aligned}\right\} \tag{5.70}$$

for $v_X \in X$ and $v_k \in Y_k$, $k = 1, \ldots, \ell$. As before, in order to reduce the cost of solving the full system, the coupling terms are neglected and the new error estimator is defined to be

$$|||\bar{e}||| \approx |||e||| \tag{5.71}$$

where

$$\bar{e} = \bar{e}_X + \bar{e}_1 + \cdots + \bar{e}_\ell \tag{5.72}$$

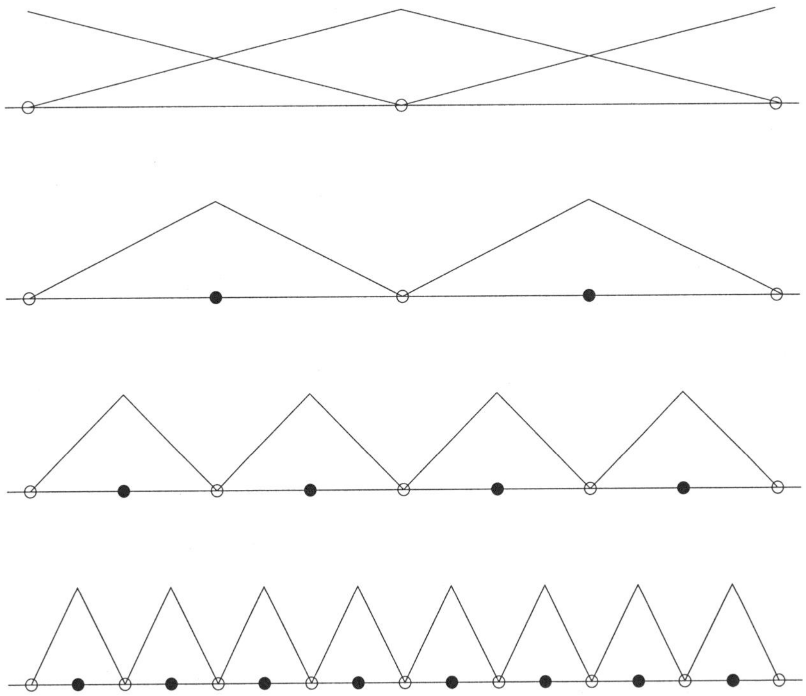

Fig. 5.6 Hierarchic basis functions for a uniform mesh in one dimension. The basis functions at each level are associated with the nodes (indicated by •) introduced by the refinement of the previous mesh.

and $\bar{e}_X \in X$, $\bar{e}_k \in Y_k$, $k = 1, \ldots, \ell$, satisfy

$$\left.\begin{array}{rcl} B(\bar{e}_X, v_X) & = & 0 \\ B(\bar{e}_1, v_1) & = & (f, v_1) - B(u_X, v_1) \\ & \vdots & \\ B(\bar{e}_\ell, v_\ell) & = & (f, v_\ell) - B(u_X, v_\ell) \end{array}\right\} \tag{5.73}$$

for $v_X \in X$ and $v_k \in Y_k$, $k = 1, \ldots, \ell$. The effect of neglecting the coupling terms is highly dependent on the selection of the basis for the spaces Y_k, and it is essential that the basis be chosen as above.

The effectiveness of the estimator may be quantified employing similar techniques as in the two-level ($\ell = 1$) case. Specifically, each element $v \in X_\ell^*$ may be uniquely decomposed in the form

$$v = v_X + v_1 + v_2 + \cdots + v_\ell \tag{5.74}$$

where $v_X \in X$ and $v_k \in Y_k$, $k = 1, \ldots, \ell$. Therefore, the operators $\Pi_k : X_\ell^* \to X_k^*$ given by the rule

$$\Pi_k v = \begin{cases} v_X + v_1 + v_2 + \cdots + v_k & \text{for } k \geq 1 \\ v_X & \text{for } k = 0 \end{cases} \tag{5.75}$$

are well-defined. Moreover, it is easy to verify that the operators are linear and are, in fact, projections. The bilinear form $D : X_\ell^* \times X_\ell^* \to \mathbb{R}$ is defined by the rule

$$D(v, w) = B(\Pi_0 v, \Pi_0 w) + \sum_{k=1}^{\ell} B\left((\Pi_k - \Pi_{k-1})v, (\Pi_k - \Pi_{k-1})w\right). \tag{5.76}$$

Now, applying the definitions gives

$$\Pi_k \bar{e} = \begin{cases} \bar{e}_X + \bar{e}_1 + \bar{e}_2 + \cdots + \bar{e}_k & \text{for } k \geq 1 \\ \bar{e}_X & \text{for } k = 0 \end{cases} \tag{5.77}$$

and consequently

$$(\Pi_k - \Pi_{k-1})\bar{e} = \bar{e}_k. \tag{5.78}$$

Thus, for $v \in X_\ell^*$,

$$\begin{aligned} D(\bar{e}, v) &= B(\bar{e}_X, v_X) + \sum_{k=1}^{\ell} B(\bar{e}_k, v_k) \\ &= B(e^*, v_X) + \sum_{k=1}^{\ell} B(e^*, v_k) \\ &= B(e^*, v), \end{aligned}$$

so the error estimators are related by the equation

$$D(\bar{e}, v) = B(e^*, v) \quad \forall v \in X_\ell^*. \tag{5.79}$$

Suppose that positive quantities λ and Λ exist such that

$$\lambda B(v, v) \leq D(v, v) \leq \Lambda B(v, v) \quad \forall v \in X_\ell^*; \tag{5.80}$$

then, by analogy with Theorem 5.3, the estimators satisfy

$$\lambda \left\| \bar{e} \right\| \leq \left\| e^* \right\| \leq \Lambda \left\| \bar{e} \right\|. \tag{5.81}$$

If the problem is posed on a domain Ω in two dimensions, then the norm equivalence

$$c \left\| v \right\|^2 \leq \left\| v_X \right\|^2 + \sum_{k=1}^{l} \left\| (\Pi_k - \Pi_{k-1}) v \right\|^2 = D(v, v) \leq C\ell^2 \left\| v \right\|^2 \tag{5.82}$$

is known [41, 117] to hold for certain positive constants c and C that are independent of ℓ and v. As a consequence, it follows that

$$c \left\| \bar{e} \right\| \leq \left\| e^* \right\| \leq C\ell^2 \left\| \bar{e} \right\|; \tag{5.83}$$

hence, the estimator is equivalent to the true error

$$c \left\| \bar{e} \right\| \leq \left\| e \right\| \leq \frac{C\ell^2}{\sqrt{1 - \beta_\ell^2}} \left\| \bar{e} \right\|. \tag{5.84}$$

Various features of this result are somewhat unsatisfactory. Firstly, the values of the constants c and C really are unknown, so the quantity $\left\| \bar{e} \right\|$ is more accurately described as an *error indicator* as opposed to an *estimator*. Furthermore, while it may be ensured that the value of constant β_ℓ is less than unity by increasing ℓ, the presence of the factor ℓ^2 in the numerator suggests that the efficiency of the indicator degenerates as ℓ grows. Nevertheless, it is worthwhile to pursue this technique further.

The solution of the system (5.73) is considerably simplified by replacing the bilinear forms appearing on the left-hand side by approximate forms that are more easily inverted. It is necessary to introduce some further notation at this stage. Let $\mathcal{N}_k$ be an indexing set for the nodes $\{\boldsymbol{x}_n^{(k)}\}$ in the mesh produced by k successive refinements of the original mesh used to construct the finite element subspace X. Let $\{\phi_m^{(k)} : m \in \mathcal{N}_k\}$ denote the standard nodal basis functions for the space X_k^*, so that

$$\phi_m^{(k)}(\boldsymbol{x}_n^{(k)}) = \delta_{mn} \quad \forall m, n \in \mathcal{N}_k. \tag{5.85}$$

The nodes generated by the kth successive subdivision of the original mesh are indexed by the set $\mathcal{N}_k \backslash \mathcal{N}_{k-1}$, and the hierarchic basis for the space Y_k is given by

$$Y_k = \text{span}\left\{ \phi_m^{(k)} : m \in \mathcal{N}_k \backslash \mathcal{N}_{k-1} \right\}. \tag{5.86}$$

The solution of the kth equation in the system (5.73) is approximated by seeking $\hat{e}_k \approx \overline{e}_k \in Y_k$, where

$$\hat{e}_k = \sum_{m \in \mathcal{N}_k \backslash \mathcal{N}_{k-1}} \hat{e}_m^{(k)} \phi_m^{(k)}, \tag{5.87}$$

and the constants $\{\hat{e}_m^{(k)}\}$ are determined by the conditions

$$B(\phi_m^{(k)}, \phi_m^{(k)}) \hat{e}_m^{(k)} = (f, \phi_m^{(k)}) - B(u_X, \phi_m^{(k)}) = (r, \phi_m^{(k)}). \tag{5.88}$$

These conditions result from ignoring the couplings between the different basis functions in the system of equations determining $\overline{e}_k$. Then, we define the error indicator to be

$$\eta^2 = \sum_{k=1}^{\ell} \sum_{m \in \mathcal{N}_k \backslash \mathcal{N}_{k-1}} \hat{e}_m^{(k)2} \left\| \phi_m^{(k)} \right\|^2 = \sum_{k=1}^{\ell} \sum_{m \in \mathcal{N}_k \backslash \mathcal{N}_{k-1}} \frac{(r, \phi^{(k)})^2}{\left\| \phi_m^{(k)} \right\|^2}. \tag{5.89}$$

It is known that the matrices here are spectrally equivalent, so that the following bounds hold:

$$c\eta \leq \|e\| \leq \frac{C\ell^2}{\sqrt{1 - \beta_\ell^2}} \eta. \tag{5.90}$$

The indicator still suffers from the same difficulties as the previous one. However, if the definition of the indicator is modified slightly to become

$$\hat{\eta}^2 = \sum_{k=1}^{\ell} \sum_{m \in \mathcal{N}_k} \frac{(r, \phi^{(k)})^2}{\left\| \phi_m^{(k)} \right\|^2}, \tag{5.91}$$

then the following equivalence holds:

$$c\hat{\eta} \leq \|e\| \leq \frac{C}{\sqrt{1 - \beta_\ell^2}} \hat{\eta} \tag{5.92}$$

where the constants c and C are bounded independently of the number of refinement levels ℓ. Therefore, letting $\ell \to \infty$, it follows that the indicator

$$\hat{\eta}_\infty^2 = \sum_{k=1}^{\infty} \sum_{m \in \mathcal{N}_k} \frac{(r, \phi^{(k)})^2}{\left\| \phi_m^{(k)} \right\|^2} \tag{5.93}$$

satisfies

$$c\hat{\eta}_\infty \leq \|e\| \leq C\hat{\eta}_\infty. \tag{5.94}$$

The presence of an infinite number of refinement levels seems to mean that the indicator has no practical value. However, this type of error indicator is the basis of an integrated algorithm that combines adaptive refinements and multilevel iterative solvers. The interested reader is referred to the monograph of Rüde [100] for further details.

5.8 BIBLIOGRAPHICAL REMARKS

The use of hierarchical bases as a vehicle for producing *a posteriori* error estimators was presented in Bank and Smith [36], and in the survey article of Bank [35]. The saturation assumption in the form presented in the text was given in the work of Bank and Weiser [37], and Bank and Smith [36]. The argument presented showing that the saturation assumption will fail to be true in general is taken form the work of Bornemann *et al.* [42].

The strengthened Cauchy–Schwarz inequality (5.26) made its debut in the analysis of multilevel iterative solution techniques, as described, for example in the review article [63]. The strengthened Cauchy–Schwarz constant for higher-order elements is studied in reference [108] and in reference [1] in the context of linear elasticity.

6
The Equilibrated Residual Method

6.1 INTRODUCTION

Implicit *a posteriori* error estimators based on the solution of local problems were discussed in Chapter 3. In particular, the element residual method entails the solution of a local Neumann boundary value problem posed over each of the elements. The data for the boundary conditions is generally obtained by averaging the boundary fluxes obtained by sampling the finite element approximation directly. However, this basic formulation of the element residual problem is unsatisfactory in various ways, not least of which is that the data for the local Neumann problem are generally incompatible with the interior residuals, with the net result that the local problem has no solution. In practice, this "difficulty" is side-stepped by reformulating the local problem over a reduced space of admissible functions where the zero energy modes have been factored out, leading to a well-posed problem.

In Chapter 3, such estimators were shown to perform rather well in certain circumstances, but evidence was also presented demonstrating that performance can be highly sensitive to the particular method used to remove the zero energy, or rigid body, modes from the space of admissible functions. The conclusion to be drawn is that the estimators suffer from unpleasant side effects arising from the incompatibility of the boundary fluxes used in the formulation of the local error residual problem.

The construction for the boundary data based on a simple averaging of boundary fluxes on neighboring elements is somewhat *ad hoc* and fails to respect the basic requirement for the local problem to be well-posed. An obvious modification is to only allow boundary fluxes leading to a well-posed local error residual problem where the boundary fluxes are in equilibrium with the interior residual loads.

In addition to equilibrium, physical considerations suggest that the boundary fluxes should be selected to satisfy certain further conditions. It will transpire that under these conditions, the resulting *a posteriori* error estimator will provide *a guaranteed upper bound* on the actual error, provided that the local problems are solved exactly. Naturally, while this is highly desirable from both a theoretical and practical viewpoint, it remains to be seen whether it is possible to develop efficient and practical algorithms for constructing such *equilibrated fluxes.*

6.2 THE EQUILIBRATED RESIDUAL METHOD

For definiteness, consider the usual model problem described in Section 1.4 and suppose that $X \subset V$ is a finite element subspace constructed on a regular partitioning $\mathcal{P}$ of the domain Ω into triangular and quadrilateral elements. The finite element approximation of this problem consists of finding $u_X \in X$ such that

$$B(u_X, v_X) = L(v_X) \quad \forall v_X \in X, \tag{6.1}$$

so that the error $e = u - u_X$ belongs to the space V and satisfies

$$B(e, v) = B(u, v) - B(u_X, v) = L(v) - B(u_X, v) \quad \forall v \in V. \tag{6.2}$$

The error measured in the energy norm may be characterized as

$$|||e||| = \sup_{0 \neq v \in V} \frac{|B(e, v)|}{|||v|||} \tag{6.3}$$

or, equally well,

$$|||e||| = \sup_{0 \neq v \in V} \frac{|L(v) - B(u_X, v)|}{|||v|||}. \tag{6.4}$$

The goal is now to decompose the residual functional appearing in this statement into contributions from the individual elements. Let $\{g_K : K \in \mathcal{P}\}$ be a set of boundary fluxes on the elements that notionally approximate the actual flux of the true solution on the element boundaries

$$g_K \approx \boldsymbol{n}_K \cdot \boldsymbol{\nabla} u|_K \text{ on } \partial K. \tag{6.5}$$

Later, it will be shown that if the approximate fluxes g_K were to coincide with the true fluxes, then the resulting error estimator would be exact.

The trace of the true fluxes is continuous on the interelement boundaries,

$$\boldsymbol{n}_K \cdot \boldsymbol{\nabla} u|_K + \boldsymbol{n}_{K'} \cdot \boldsymbol{\nabla} u|_{K'} = 0 \text{ on } \partial K \cap \partial K', \tag{6.6}$$

and so, by analogy, the approximate fluxes are required to satisfy the condition

$$g_K + g_{K'} = 0 \text{ on } \partial K \cap \partial K'. \tag{6.7}$$

Physically, this condition expresses the requirement that flux should not be generated on the actual interface. Naturally, since the true flux is known on the portion of the exterior boundary Γ_N where a Neumann condition is prescribed, the approximate flux is chosen to coincide with the true flux on Γ_N,

$$g_K = g \text{ on } \partial K \cap \Gamma_N. \tag{6.8}$$

Together, the conditions (6.7) and (6.8) imply that for all $v \in V$,

$$\int_{\Gamma_N} gv \,\mathrm{d}s = \sum_{K \in \mathcal{P}} \int_{\partial K} g_K v \,\mathrm{d}s. \tag{6.9}$$

This is easily seen since, thanks to (6.7), the contributions from the interior edges cancel pairwise, while the exterior edges are dealt with using (6.8) and by noting the fact that v vanishes on the portion of the boundary where a Dirichlet condition is imposed.

The error residual functional may be decomposed into contributions from the individual elements

$$L(v) - B(u_X, v) = \sum_{K \in \mathcal{P}} \{(f, v)_K - B_K(u_X, v)\} + \int_{\Gamma_N} gv \,\mathrm{d}s \tag{6.10}$$

for each $v \in V$, where

$$B_K(u, v) = \int_K (\nabla u \cdot \nabla v + cuv) \,\mathrm{d}\boldsymbol{x} \tag{6.11}$$

and

$$(f, v)_K = \int_K fv \,\mathrm{d}\boldsymbol{x}. \tag{6.12}$$

As a consequence of (6.9), we deduce that for all $v \in V$,

$$L(v) - B(u_X, v) = \sum_{K \in \mathcal{P}} \left\{(f, v)_K - B_K(u_X, v) + \int_{\partial K} g_K v \,\mathrm{d}s\right\}. \tag{6.13}$$

The term in parentheses is a local residual functional over element K. This may be represented in terms of the solution $\phi_K \in V_K$ of the local residual problem

$$B_K(\phi_K, v) = (f, v)_K - B_K(u_X, v) + \int_{\partial K} g_K v \,\mathrm{d}s \quad \forall v \in V_K \tag{6.14}$$

where V_K is the space comprised of the locally admissible functions,

$$V_K = \left\{v \in H^1(K) : v = 0 \text{ on } \Gamma_D \cap \partial K\right\}. \tag{6.15}$$

Here, it is tacitly assumed that the local problem has a solution ϕ_K. The solution always exists and is unique whenever the coefficient c appearing in the bilinear form is positive. However, if the absolute term vanishes, then the problem will have a solution if and only if the boundary data and interior residual satisfy the compatibility or *equilibration condition*

$$0 = (f, 1)_K - B_K(u_X, 1) + \int_{\partial K} g_K \, \mathrm{d}s \tag{6.16}$$

whenever $v \in V_K$ may be chosen as the constant function $v(\boldsymbol{x}) \equiv 1$. This condition expresses the fact that the boundary flux g_K must be in equilibrium with the interior load.

By inserting (6.14) into (6.13), it follows that for all $v \in V$,

$$B(e, v) = L(v) - B(u_X, v) = \sum_{K \in \mathcal{P}} B_K(\phi_K, v). \tag{6.17}$$

This is an extremely useful fact, showing that the global error residual may be decomposed into the sum of local residuals.

One immediate consequence of this result is the upper bound on the true error asserted in the introduction. This follows from the Cauchy–Schwarz inequality,

$$|B(e, v)| \leq \sum_{K \in \mathcal{P}} \|\phi_K\|_K \, \|v\|_K \tag{6.18}$$

and then again thanks to the Cauchy–Schwarz inequality we have

$$|B(e, v)| \leq \left\{ \sum_{K \in \mathcal{P}} \|\phi_K\|_K^2 \right\}^{1/2} \|v\| \, . \tag{6.19}$$

Finally, the characterization (6.3) leads to the conclusion

$$\|e\|^2 \leq \sum_{K \in \mathcal{P}} \|\phi_K\|_K^2 \, . \tag{6.20}$$

These developments are recorded in the following theorem:

Theorem 6.1 *Let $\{g_K : K \in \mathcal{P}\}$ be any set of boundary fluxes satisfying conditions (6.7) and (6.8). In addition, if the absolute term c vanishes, then it is assumed that the fluxes satisfy the equilibration condition (6.16) on all elements that do not abut the Dirichlet boundary Γ_D. Then, the global error residual may be decomposed into local contributions*

$$B(e, v) = L(v) - B(u_X, v) = \sum_{K \in \mathcal{P}} B_K(\phi_K, v) \quad v \in V \tag{6.21}$$

where $\phi_K \in V_K$ is the solution of the local problem (6.14). The global error in the finite element approximation may be bounded by

$$\|e\|^2 \leq \sum_{K \in \mathcal{P}} \|\phi_K\|_K^2 \, . \tag{6.22}$$

Proof. The proof follows at once from the previous arguments. ∎

The significance of Theorem 6.1 is that one can obtain guaranteed upper bounds on the error measured in the energy norm under certain conditions on the local fluxes $\{g_K\}$. Naturally, the *quality* of the upper bound will depend on how accurately the local fluxes approximate the true fluxes. In fact, if it were possible to select g_K to be the true flux, then the upper bound would coincide with the true error. This follows at once from observing that by taking g_K equal to the true flux, the data for the local problem (6.14) may be rewritten using integration by parts as

$$(f, v)_K - B_K(u_X, v) + \int_{\partial K} \boldsymbol{n}_K \cdot \boldsymbol{\nabla} u \, \mathrm{d}s = B_K(u, v) - B_K(u_X, v)$$

and consequently the solution of the local problem ϕ_K coincides with the true error e (possibly up to the addition of a rigid body motion). In this eventuality, the estimated error would *equal* the true error.

It is clear that the equilibration condition (6.16) is essential for the well-posedness of the local problems in the case when the absolute term c is absent. However, one must be wary of hastily drawing the conclusion that the condition plays no role if the absolute term is nonzero, as is shown by the following example.

The energy norm $\|\phi_K\|_K$ of the solution of the local problem may be characterized as

$$\|\phi_K\|_K = \sup_{0 \neq v \in V_K} \frac{|B_K(\phi_K, v)|}{\|v\|_K}. \tag{6.23}$$

Suppose that an element K does not abut the Dirichlet boundary Γ_D, meaning that the constant function $v = 1$ belongs to V_K. Then, selecting v to be the constant function in (6.23) gives the lower bound

$$\|\phi_K\|_K \geq \frac{|B_K(\phi_K, 1)|}{\|1\|_K}. \tag{6.24}$$

The numerator may be rewritten with the help of (6.14) as

$$B_K(\phi_K, 1) = (f, 1)_K - B_K(u_X, 1) + \int_{\partial K} g_K \, \mathrm{d}s \tag{6.25}$$

while

$$\|1\|_K^2 = \int_K c \, \mathrm{d}\boldsymbol{x}, \tag{6.26}$$

and hence

$$\|\phi_K\|_K^2 \geq \frac{|(f, 1)_K - B_K(u_X, 1) + \int_{\partial K} g_K \, \mathrm{d}s|^2}{\int_K c \, \mathrm{d}\boldsymbol{x}}. \tag{6.27}$$

If the equilibration condition (6.16) is violated, then the numerator will be nonzero whilst the denominator is of order ch_K^2. This means that there is a danger that the energy of the local solution ϕ_K might blow up as the mesh is refined, unless the quantity appearing in the numerator decays at the same rate as $h_K \to 0$. The rate of decay that is required to prevent blow-up practically means that the numerator has to vanish; in other words, the equilibration condition must be satisfied.

The moral of this story is that the equilibration condition, while not strictly necessary for the well-posedness of the local problems when the absolute term c is present, still has a direct impact on the *quality* of the upper bounds and, in any event, must be satisfied in the limit as the mesh size tends to zero if one is to obtain sharp bounds.

6.3 THE EQUILIBRATED FLUX CONDITIONS

The previous discussion indicates the desirability of constructing approximate boundary fluxes $\{g_K\}$ satisfying the *zeroth-order equilibration conditions*:

$$\left.\begin{aligned} (f,1)_K - B_K(u_X,1) + \int_{\partial K} g_K \,\mathrm{d}s &= 0 \\ g_K + g_{K'} &= 0 \text{ on } \partial K \cap \partial K' \\ g_K &= g \text{ on } \partial K \cap \Gamma_N. \end{aligned}\right\} \qquad (6.28)$$

The terminology *zeroth-order* simply refers to the fact that the fluxes are compatible with respect to constants and is not to be confused with an order of convergence. Later, the notion of *p th-order equilibration* will be defined.

There are now two main issues to be resolved:

- Is it possible for fluxes $\{g_K\}$ to be reconstructed from the finite element approximation u_X and the data such that the equilibration conditions (6.28) are satisfied?

- If so, is it possible for this to be achieved without having to resort to a global computation?

Perhaps surprisingly, we shall later show that the answer to both of these questions is affirmative.

One fact worth recording at the outset is that *the boundary fluxes are not uniquely determined by the conditions (6.28).* For instance, suppose the bilinear form $B(\cdot,\cdot)$ corresponds to the Laplace operator and let χ be *any* harmonic function on Ω such that $\partial\chi/\partial n$ vanishes on the Neumann boundary Γ_N. The freedom in the choice of χ on the Dirichlet boundary Γ_D means that there are infinitely many possibilities for χ. Now, given a set of fluxes $\{g_K\}$ satisfying the conditions (6.28), it is simple to verify that the fluxes $\{g_K^\chi\}$ defined by

$$g_K^\chi = g_K + \boldsymbol{n}_K \cdot \boldsymbol{\nabla}\chi \text{ on } \partial K$$

also satisfy the conditions (6.28). This example indicates that the problem of selecting equilibrated boundary fluxes is underdetermined, and it is only to be expected that this lack of uniqueness will manifest itself later.

The rationale behind the second question is that in order for the resulting error estimator to be economical, it is essential to reduce the computation of equilibrated fluxes to independent local computations. It is worth observing that the equilibration conditions cannot be satisfied working with a single element in isolation, since the flux approximation g_K is coupled to the flux $g_{K'}$ on neighboring elements by the consistency condition,

$$g_K + g_{K'} = 0 \text{ on } \partial K \cap \partial K'.$$

In effect, while the local problems (6.14) leading to the error estimator are apparently decoupled, the coupling persists indirectly through the constraints on the boundary fluxes. Realizing that one cannot hope to work over a single element in isolation will naturally lead to the question of whether it is possible to reach a compromise and reconstruct equilibrated fluxes by dealing with *local patches* of elements (as opposed to the whole mesh)—and this time the answer is positive as will be shown in the next section.

6.4 EQUILIBRATED FLUXES ON REGULAR PARTITIONS

This section is devoted to describing a procedure for constructing sets of boundary fluxes satisfying the zeroth-order equilibration conditions (6.28).

In the interest of clarity, it will first be assumed that the finite element subspace X is constructed using first-order (linear or bilinear) elements on a regular partitioning $\mathcal{P}$ of the domain Ω into triangular or quadrilateral elements. As usual, the finite element approximation is defined by the condition $u_X \in X$:

$$B(u_X, v) = L(v) = (f, v) + \int_{\Gamma_N} gv \, ds \quad \forall v \in X. \tag{6.29}$$

Let $\mathcal{N}$ denote the nodes in the space X and let $\{\theta_n : n \in \mathcal{N}\}$ be a Lagrange basis for the space X, as described in Chapter 1, characterized by the conditions

$$\theta_m(\boldsymbol{x}_n) = \delta_{mn} \quad m, n \in \mathcal{N}$$

so that

$$\sum_{n \in \mathcal{N}} \theta_n(\boldsymbol{x}) = 1 \text{ in } \Omega.$$

The nodes on an element K are denoted by $\mathcal{N}(K)$, and it follows that the Lagrange basis functions on the element satisfy

$$\sum_{n \in \mathcal{N}(K)} \theta_n(\boldsymbol{x}) = 1 \text{ in } K. \tag{6.30}$$

Similarly, the nodes on an edge γ are denoted by $\mathcal{N}(\gamma)$ and the basis functions satisfy

$$\sum_{n\in\mathcal{N}(\gamma)} \theta_n(\boldsymbol{x}) = 1 \text{ on } \gamma. \tag{6.31}$$

6.4.1 First-Order Equilibration Condition

The procedure that will be developed produces sets of fluxes $\{g_K\}$ that satisfy the *first-order equilibration conditions*:

$$\left.\begin{aligned} (f,\theta_n)_K - B_K(u_X,\theta_n) + \textstyle\int_{\partial K} g_K\theta_n\,\mathrm{ds} &= 0 \quad \forall n \in \mathcal{N}(K) \\ g_K + g_{K'} &= 0 \text{ on } \partial K \cap \partial K' \\ g_K &= g \text{ on } \partial K \cap \Gamma_N. \end{aligned}\right\} \tag{6.32}$$

The terminology *first-order* reflects that the first condition is required to hold for first-order finite element functions, as opposed to simply constant functions as in case of the zeroth-order condition (6.28). This condition imposes stricter requirements on the fluxes than the zeroth-order equilibration condition. Indeed, the zeroth-order condition is a direct consequence of (6.32) thanks to the fact that the constant function $v = 1$ may be written as a sum of first-order basis functions as in (6.30).

6.4.2 The Form of the Boundary Fluxes

The final condition of (6.32) completely determines the form of the recovered flux g_K on a single edge γ of the portion Γ_N of the boundary where Neumann data are prescribed. However, on the remaining edges some latitude is possible, and it will be found convenient if the boundary flux is taken to be a *linear function*, thus:

$$g_K|_\gamma \begin{cases} \text{is the Neumann data } g \text{ on } \gamma \subset \Gamma_N, \text{ or} \\ \text{belongs to } \operatorname{span}\{\theta_n : n \in \mathcal{N}(\gamma)\} \text{ on remaining edges.} \end{cases} \tag{6.33}$$

Having settled on this form for the fluxes, it remains to specify two independent degrees of freedom in order to characterize the linear function uniquely. It is at this stage we make a key decision: *The degrees of freedom for the flux $g_K|_\gamma$ on edge γ are chosen to be the moments of the flux weighted against the basis functions on edge γ:*

$$\mu^\gamma_{K,n} = \int_\gamma g_K\theta_n\,\mathrm{ds}, \quad n \in \mathcal{N}(\gamma). \tag{6.34}$$

This choice for the degrees of freedom plays a pivotal role in avoiding having to deal with a global problem, by reducing the determination of the fluxes to computations over small, local patches of elements.

It is a relatively simple matter to reconstruct the actual flux from the moments once they are in hand. For instance, suppose that $\mathcal{N}(\gamma) = \{\ell, r\}$ so that the endpoints of the edge are located at the nodes $\boldsymbol{x}_\ell$ and $\boldsymbol{x}_r$, and write

$$g_K|_\gamma = \alpha_\ell \theta_\ell + \alpha_r \theta_r$$

where α_ℓ and α_r are constants to be determined in terms of the moments of the flux. It follows from definition (6.34) that the unknowns satisfy the conditions

$$\begin{aligned} (\theta_\ell, \theta_\ell)_\gamma \alpha_\ell + (\theta_\ell, \theta_r)_\gamma \alpha_r &= \mu_{K,\ell}^\gamma \\ (\theta_r, \theta_\ell)_\gamma \alpha_\ell + (\theta_r, \theta_r)_\gamma \alpha_r &= \mu_{K,r}^\gamma \end{aligned}$$

where $(\cdot, \cdot)_\gamma$ denotes the inner product on $L_2(\gamma)$. Equally well,

$$\boldsymbol{M}_\gamma \begin{bmatrix} \alpha_\ell \\ \alpha_r \end{bmatrix} = \begin{bmatrix} \mu_{K,\ell}^\gamma \\ \mu_{K,r}^\gamma \end{bmatrix} \tag{6.35}$$

where $\boldsymbol{M}_\gamma$ is the *mass matrix* for the basis functions on the edge γ. A simple computation shows that if the length of the edge is h, then

$$\boldsymbol{M}_\gamma = \frac{h}{6} \begin{bmatrix} 2 & 1 \\ 1 & 2 \end{bmatrix}$$

and hence

$$\alpha_\ell = \frac{2}{h}\left(2\mu_{K,\ell}^\gamma - \mu_{K,r}^\gamma\right); \quad \alpha_r = \frac{2}{h}\left(-\mu_{K,\ell}^\gamma + 2\mu_{K,r}^\gamma\right).$$

The actual flux is given directly in terms of the moments by the expression

$$g_K|_\gamma = \frac{2}{h}\left\{(2\mu_{K,\ell}^\gamma - \mu_{K,r}^\gamma)\theta_\ell + (-\mu_{K,\ell}^\gamma + 2\mu_{K,r}^\gamma)\theta_r\right\}. \tag{6.36}$$

The relationship between this form of the recovered flux and the one proposed in reference [77] may be exhibited by noting that (6.36) could be rewritten in the form

$$g_K|_\gamma = \mu_{K,\ell}^\gamma \psi_\ell + \mu_{K,r}^\gamma \psi_r$$

where ψ_ℓ and ψ_r are the piecewise linear functions

$$\psi_\ell = \frac{2}{h}(2\theta_\ell - \theta_r); \quad \psi_r = \frac{2}{h}(-\theta_\ell + 2\theta_r).$$

This form for the flux was proposed in reference [77].

6.4.3 Equilibration Conditions in Terms of the Moments

In order to determine the boundary fluxes, it is sufficient to determine the moments of the flux with respect to the basis functions. The first-order equilibration conditions (6.32) for the continuous flux g_K may be reformulated in terms of the flux moments as follows:

$$\left.\begin{array}{rcll} \sum_{\gamma\subset\partial K}\mu^{\gamma}_{K,n} &=& \Delta_K(\theta_n) & \forall n\in\mathcal{N}(K) \\ \mu^{\gamma}_{K,n}+\mu^{\gamma}_{K',n} &=& 0 & \forall n\in\mathcal{N}(\gamma),\quad \gamma=\partial K\cap\partial K' \\ \mu^{\gamma}_{K,n} &=& \int_\gamma g\theta_n\,\mathrm{d}s & \forall n\in\mathcal{N}(\gamma),\quad \gamma=\partial K\cap\Gamma_N \end{array}\right\} \tag{6.37}$$

where

$$\Delta_K(\theta_n)=B_K(u_X,\theta_n)-(f,\theta_n)_K. \tag{6.38}$$

The advantage of selecting the flux moments to be the degrees of freedom for the fluxes is now evident: *The conditions (6.37) represent a set of independent conditions associated with each node $n\in\mathcal{N}$.* In terms of the practical implementation, this means that the computations associated with each node n may be performed independently and concurrently. Furthermore, the domain influenced by a particular basis function θ_n is comprised of *small, local patches* of elements and their edges. This decoupling into local problems avoids the obligation to treat a globally coupled problem in keeping with our original requirements.

6.4.4 Local Patch Problems for the Flux Moments

Let $n\in\mathcal{N}$ be any node and let θ_n be the Lagrange basis function associated with the node. The set of elements $\mathcal{P}_n$ influenced by this basis function is defined by

$$\mathcal{P}_n=\{K\in\mathcal{P}:n\in\mathcal{N}(K)\}, \tag{6.39}$$

or, simply stated, $\mathcal{P}_n$ is the patch of elements with a vertex at $\boldsymbol{x}_n$. Similarly, the set of edges influenced by the basis function is

$$\mathcal{E}_n=\{\gamma\in\partial\mathcal{P}:n\in\mathcal{N}(\gamma)\} \tag{6.40}$$

and consists of those edges having a vertex at $\boldsymbol{x}_n$.

The conditions (6.37) take one of four distinct structures depending on the location of the node $\boldsymbol{x}_n$ and the form of the boundary conditions applied in the neighborhood of the nodes.

1. Interior Vertex: The elements and edges are labeled as shown in Figure 6.1. The moment equilibration conditions (6.37) for the elements $K\in\mathcal{P}_n$ associ-

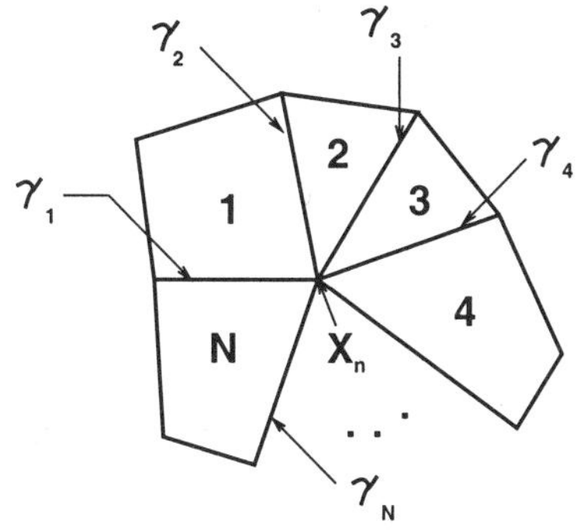

Fig. 6.1 The patches $\mathcal{P}_n$ and $\mathcal{E}_n$ of elements and edges influenced by the basis function θ_n associated with an interior vertex located at $\boldsymbol{x}_n$.

ated with the node n assume the form

$$\left.\begin{aligned} \mu_{1,n}^{\gamma_1} + \mu_{1,n}^{\gamma_2} &= \Delta_1(\theta_n) \\ \mu_{2,n}^{\gamma_2} + \mu_{2,n}^{\gamma_3} &= \Delta_2(\theta_n) \\ &\vdots \\ \mu_{N,n}^{\gamma_N} + \mu_{N,n}^{\gamma_1} &= \Delta_N(\theta_n) \end{aligned}\right\}$$

with constraints on the *interior* edges

$$\left.\begin{aligned} \mu_{1,n}^{\gamma_1} + \mu_{N,n}^{\gamma_1} &= 0 \\ \mu_{2,n}^{\gamma_2} + \mu_{1,n}^{\gamma_2} &= 0 \\ &\vdots \\ \mu_{N,n}^{\gamma_N} + \mu_{N-1,n}^{\gamma_N} &= 0. \end{aligned}\right\}$$

The constraint in (6.37) concerning edges on the Neumann boundary are, of course, vacuous in the case of an interior vertex.

2. Boundary Vertex: The elements and edges are labeled as in Figure 6.2. The moment equilibration conditions (6.37) become

$$\left.\begin{aligned} \mu_{1,n}^{\gamma_1} + \mu_{1,n}^{\gamma_2} &= \Delta_1(\theta_n) \\ \mu_{2,n}^{\gamma_2} + \mu_{2,n}^{\gamma_3} &= \Delta_2(\theta_n) \\ &\vdots \\ \mu_{N,n}^{\gamma_N} + \mu_{N,n}^{\gamma_{N+1}} &= \Delta_N(\theta_n) \end{aligned}\right\}$$

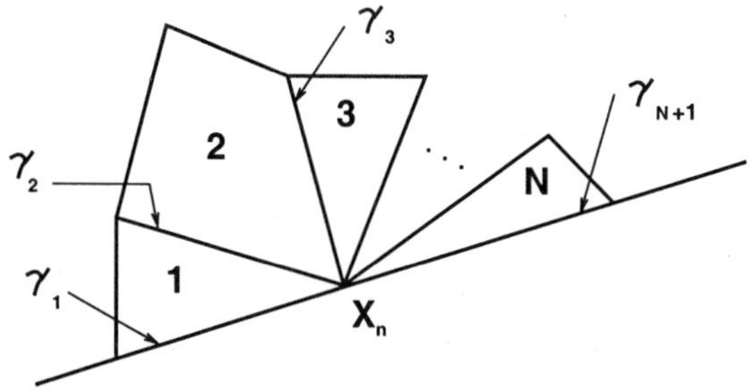

Fig. 6.2 The patches $\mathcal{P}_n$ and $\mathcal{E}_n$ of elements and edges influenced by the basis function θ_n associated with a node located on the boundary of the domain.

with the following constraints on the *interior* edges

$$\left.\begin{aligned} \mu_{2,n}^{\gamma_2} + \mu_{1,n}^{\gamma_2} &= 0 \\ \mu_{3,n}^{\gamma_3} + \mu_{2,n}^{\gamma_3} &= 0 \\ &\vdots \\ \mu_{N,n}^{\gamma_N} + \mu_{N-1,n}^{\gamma_N} &= 0. \end{aligned}\right\}$$

The *exterior* edges γ_1 and γ_{N+1} are treated differently depending on the type of boundary condition (Neumann or Dirichlet) applied on the exterior boundary. If an edge lies on the portion Γ_N of the boundary where a Neumann condition is prescribed, then the final constraint in (6.37) is active. Conversely, if the edge lies on the portion Γ_D where a Dirichlet condition is applied, then there are no constraints on the flux moment.

There are three distinct possibilities:

2(a) Neumann–Neumann: If $\gamma_1 \subset \Gamma_N$ and $\gamma_{N+1} \subset \Gamma_N$, then

$$\begin{aligned} \mu_{1,n}^{\gamma_1} &= \int_{\gamma_1} g\theta_n \, \mathrm{d}s \\ \mu_{N,n}^{\gamma_{N+1}} &= \int_{\gamma_{N+1}} g\theta_n \, \mathrm{d}s. \end{aligned}$$

2(b) Dirichlet–Neumann: If $\gamma_1 \subset \Gamma_N$ and $\gamma_{N+1} \subset \Gamma_D$, then

$$\begin{aligned} \mu_{1,n}^{\gamma_1} &= \int_{\gamma_1} g\theta_n \, \mathrm{d}s \\ \mu_{N,n}^{\gamma_{N+1}} &= \text{unconstrained}. \end{aligned}$$

Table 6.1 The various possibilities concerning the existence and uniqueness of flux moments satisfying the local patch problems comprised of N elements, for each of the four distinct cases enumerated in the text

	Unknowns		*Constraints*		
Case	Flux Moments	Element Equilibrium	Interior Edges	Exterior Edges	Solution
1	$2N$	N	N	0	Nonunique
2(a)	$2N$	N	$N-1$	2	Unique
2(b)	$2N$	N	$N-1$	1	Unique
2(c)	$2N$	N	$N-1$	0	Nonunique

2(c) Dirichlet–Dirichlet: If $\gamma_1 \subset \Gamma_D$ and $\gamma_{N+1} \subset \Gamma_D$, then

$$\begin{aligned} \mu_{1,n}^{\gamma_1} &= \text{unconstrained} \\ \mu_{N,n}^{\gamma_{N+1}} &= \text{unconstrained.} \end{aligned}$$

These four cases are the only distinct possibilities. Nonetheless, the problems exhibit quite different characteristics. For instance, in all cases there are $2N$ unknown flux moments (two in each of the N elements), and in all cases there are N constraints associated with the equilibrium condition on the elements, but here the similarity ends. The situation is summarized in Table 6.1, where it will be noted that a solution exists in all cases. The justification of these statements is deferred to the next section.

6.4.5 Procedure for Resolution of Patch Problems

The examples in the previous section show that the solution of the moment conditions (6.37) may well be nonunique. Applying Theorem 6.1 shows that any solution of the system selected at random will lead to an upper bound on the error. However, the quality of the bound is likely to be poor unless a solution is picked with some care. Earlier, it was shown that if the flux g_K coincides with the flux of the true solution,

$$g_K \approx \boldsymbol{n}_K \cdot \boldsymbol{\nabla} u \text{ on } \partial K,$$

then the upper bound would be exact. The ideal situation would be to choose the approximate flux moments such that

$$\mu_{K,n}^{\gamma} \approx \int_{\gamma} \theta_n \boldsymbol{n}_K \cdot \boldsymbol{\nabla} u \, \mathrm{d}s,$$

assuming that these quantities are well-defined. Of course, the true fluxes are generally unknown and so, in the absence of any better point of reference, the flux moments are selected so that

$$\mu_{K,n}^{\gamma} \approx \tilde{\mu}_{K,n}^{\gamma} = \int_{\gamma} \theta_n \, \boldsymbol{n}_K \cdot \nabla u_X|_K \, \mathrm{d}s. \tag{6.41}$$

The role of these conditions is to assist with the removal of any possible nonuniqueness by seeking flux moments that minimize the objective

$$\frac{1}{2} \sum_{K \in \mathcal{P}_n} \sum_{\gamma \subset \partial K} \left(\mu_{K,n}^{\gamma} - \tilde{\mu}_{K,n}^{\gamma} \right)^2 \tag{6.42}$$

subject to the equilibration conditions (6.37).

The optimality conditions for problem (6.42) may be derived through the introduction of Lagrange multipliers for each of the constraints in (6.37). The Lagrangian is given by

$$\begin{aligned} \mathcal{L}\left(\{\tilde{\mu}_{K,n}^{\gamma}\}, \{\lambda_\gamma\}, \{\sigma_K\} \right) &= \frac{1}{2} \sum_{K \in \mathcal{P}_n} \sum_{\gamma \subset \partial K} \left(\mu_{K,n}^{\gamma} - \tilde{\mu}_{K,n}^{\gamma} \right)^2 \\ &\quad + \sum_{K \in \mathcal{P}_n} \sigma_{K,n} \left(\Delta_K(\theta_n) - \sum_{\gamma \subset \partial K} \mu_{K,n}^{\gamma} \right) \\ &\quad + \sum_{\gamma = \partial K \cap \partial K'} \lambda_{\gamma,n} \left(\mu_{K,n}^{\gamma} + \mu_{K',n}^{\gamma} \right) \\ &\quad + \sum_{\gamma = \partial K \cap \Gamma_N} \lambda_{\gamma,n} \left(\mu_{K,n}^{\gamma} - \int_{\gamma} g\theta_n \, \mathrm{d}s \right). \end{aligned}$$

As a notational convenience, the value of the Lagrange multiplier $\lambda_{\gamma,n}$ on a Dirichlet edge $\gamma \subset \Gamma_D$ is set to zero, corresponding to the flux moments being unconstrained on these edges. With this convention, the Euler conditions for a stationary point are then given by (6.37) supplemented with the additional conditions

$$\mu_{K,n}^{\gamma} - \tilde{\mu}_{K,n}^{\gamma} - \sigma_{K,n} + \lambda_{\gamma,n} = 0 \tag{6.43}$$

and

$$\lambda_{\gamma,n} = 0 \text{ on } \gamma \subset \Gamma_D. \tag{6.44}$$

These conditions may be used in conjunction with the second and third parts of (6.37) to obtain the following formula for the edge multipliers:

$$\lambda_{\gamma,n} = \begin{cases} \frac{1}{2}\left(\sigma_{K,n} + \sigma_{K',n} + \tilde{\mu}_{K,n}^{\gamma} + \tilde{\mu}_{K',n}^{\gamma} \right) & \gamma = \partial K \cap \partial K' \\ \sigma_{K,n} + \tilde{\mu}_{K,n}^{\gamma} - \int_{\gamma} g\theta_n \, \mathrm{d}s & \gamma = \partial K \cap \Gamma_N \\ 0 & \gamma = \partial K \cap \Gamma_D. \end{cases}$$

If this expression is substituted back into (6.43), then one arrives at the following expression for the flux moments:

$$\mu_{K,n}^{\gamma} = \begin{cases} \frac{1}{2}\left(\sigma_{K,n} - \sigma_{K',n} + \tilde{\mu}_{K,n}^{\gamma} - \tilde{\mu}_{K',n}^{\gamma}\right) & \gamma = \partial K \cap \partial K' \\ \int_{\gamma} g\theta_n \, ds & \gamma = \partial K \cap \Gamma_N \\ \sigma_{K,n} + \tilde{\mu}_{K,n}^{\gamma} & \gamma = \partial K \cap \Gamma_D. \end{cases} \tag{6.45}$$

Finally, inserting this information into the first equation in (6.37) leads to the following set of conditions for the Lagrange multipliers $\{\sigma_{K,n} : K \in \mathcal{P}_n\}$:

$$\frac{1}{2}\sum_{\gamma=\partial K \cap \partial K'} (\sigma_{K,n} - \sigma_{K',n}) + \sum_{\gamma \subset \partial K \cap \Gamma_D} \sigma_{K,n} = \widetilde{\Delta}_K(\theta_n) \quad \forall K \in \mathcal{P}_n \tag{6.46}$$

where

$$\begin{aligned} \widetilde{\Delta}_K(\theta_n) &= \Delta_K(\theta_n) - \frac{1}{2}\sum_{\gamma=\partial K \cap \partial K'} (\tilde{\mu}_{K,n}^{\gamma} - \tilde{\mu}_{K',n}^{\gamma}) \\ &\quad - \sum_{\gamma \subset \partial K \cap \Gamma_D} \tilde{\mu}_{K,n}^{\gamma} - \sum_{\gamma \subset \partial K \cap \Gamma_N} \int_{\gamma} g\theta_n \, ds. \end{aligned} \tag{6.47}$$

This quantity may be written in the alternative form

$$\widetilde{\Delta}_K(\theta_n) = B_K(u_X, \theta_n) - (f, \theta_n)_K - \int_{\partial K} \left\langle \frac{\partial u_X}{\partial n_K} \right\rangle \theta_n \, ds \tag{6.48}$$

where we recall the notation

$$\left\langle \frac{\partial u_X}{\partial n_K} \right\rangle = \begin{cases} \frac{1}{2}\boldsymbol{n}_K \cdot \{(\boldsymbol{\nabla} u_X)_K + (\boldsymbol{\nabla} u_X)_{K'}\} & \text{on } \partial K \cap \partial K' \\ \boldsymbol{n}_K \cdot (\boldsymbol{\nabla} u_X)_K & \text{on } \partial K \cap \Gamma_D \\ g & \text{on } \partial K \cap \Gamma_N. \end{cases} \tag{6.49}$$

The conditions (6.46) represent a linear algebraic system over the element patches $\mathcal{P}_n$ with unknowns $\{\sigma_{K,n} : K \in \mathcal{P}_n\}$ corresponding to the elements in the patch. The specific form of the systems for each of the cases identified earlier is given below (where the subscript n is omitted).

1. Interior Vertex: The equations for the interior patch shown in Figure 6.1 are given by

$$\frac{1}{2}\begin{bmatrix} 2 & -1 & & \cdots & & -1 \\ -1 & 2 & -1 & \cdots & & 0 \\ \vdots & & & & & \vdots \\ 0 & & \cdots & -1 & 2 & -1 \\ -1 & & \cdots & & -1 & 2 \end{bmatrix} \begin{bmatrix} \sigma_1 \\ \sigma_2 \\ \vdots \\ \sigma_{N-1} \\ \sigma_N \end{bmatrix} = \begin{bmatrix} \widetilde{\Delta}_1(\theta_n) \\ \widetilde{\Delta}_2(\theta_n) \\ \vdots \\ \widetilde{\Delta}_{N-1}(\theta_n) \\ \widetilde{\Delta}_N(\theta_n) \end{bmatrix}.$$

2. Boundary Vertex: Consider the equations for the patch shown in Figure 6.2. As before, the equations assume various forms depending on the type of boundary conditions imposed on the exterior boundary.

2(a) Neumann–Neumann:

$$\frac{1}{2}\begin{bmatrix} 1 & -1 & & \cdots & & 0 \\ -1 & 2 & -1 & \cdots & & 0 \\ \vdots & & & & & \vdots \\ 0 & & \cdots & -1 & 2 & -1 \\ 0 & & \cdots & & -1 & 1 \end{bmatrix} \begin{bmatrix} \sigma_1 \\ \sigma_2 \\ \vdots \\ \sigma_{N-1} \\ \sigma_N \end{bmatrix} = \begin{bmatrix} \widetilde{\Delta}_1(\theta_n) \\ \widetilde{\Delta}_2(\theta_n) \\ \vdots \\ \widetilde{\Delta}_{N-1}(\theta_n) \\ \widetilde{\Delta}_N(\theta_n) \end{bmatrix}.$$

2(b) Dirichlet–Neumann:

$$\frac{1}{2}\begin{bmatrix} 3 & -1 & & \cdots & & 0 \\ -1 & 2 & -1 & \cdots & & 0 \\ \vdots & & & & & \vdots \\ 0 & & \cdots & -1 & 2 & -1 \\ 0 & & \cdots & & -1 & 1 \end{bmatrix} \begin{bmatrix} \sigma_1 \\ \sigma_2 \\ \vdots \\ \sigma_{N-1} \\ \sigma_N \end{bmatrix} = \begin{bmatrix} \widetilde{\Delta}_1(\theta_n) \\ \widetilde{\Delta}_2(\theta_n) \\ \vdots \\ \widetilde{\Delta}_{N-1}(\theta_n) \\ \widetilde{\Delta}_N(\theta_n) \end{bmatrix}.$$

2(c) Dirichlet–Dirichlet:

$$\frac{1}{2}\begin{bmatrix} 3 & -1 & & \cdots & & 0 \\ -1 & 2 & -1 & \cdots & & 0 \\ \vdots & & & & & \vdots \\ 0 & & \cdots & -1 & 2 & -1 \\ 0 & & \cdots & & -1 & 3 \end{bmatrix} \begin{bmatrix} \sigma_1 \\ \sigma_2 \\ \vdots \\ \sigma_{N-1} \\ \sigma_N \end{bmatrix} = \begin{bmatrix} \widetilde{\Delta}_1(\theta_n) \\ \widetilde{\Delta}_2(\theta_n) \\ \vdots \\ \widetilde{\Delta}_{N-1}(\theta_n) \\ \widetilde{\Delta}_N(\theta_n) \end{bmatrix}.$$

The matrices involved contain only integer entries and are easily assembled by examining the *topology* of the patch, and they were referred to as *topology matrices* in [8].

It will be observed that for an interior vertex and a Neumann–Neumann boundary vertex, the matrices are singular. This raises the issue of solvability of the systems. In both cases, the null space of the matrices is the vector $\mathbf{1}$,

$$\mathbf{1} = [1, 1, \ldots, 1]^\top$$

implying that a solution exists if and only if the sum of the components of the right-hand data vanishes. This may be confirmed as follows. Firstly, observe that in the summation of the quantities $\widetilde{\Delta}_K(\theta_n)$ over all elements in the patch $\mathcal{P}_n$, the terms involving the interior edges cancel pairwise between the two elements sharing the edge. Consequently, the sum reduces to

$$\sum_{K\in\mathcal{P}_n} \widetilde{\Delta}_K(\theta_n) = \sum_{K\in\mathcal{P}_n} \Delta_K(\theta_n) - \int_\gamma g\theta_n \, \mathrm{ds}$$

since in each of the cases of interest, the patch $\mathcal{P}_n$ does not abut the Dirichlet boundary Γ_D. The defining property (6.29) means that

$$\sum_{K \in \mathcal{P}_n} \Delta_K(\theta_n) = \sum_{\gamma = \Gamma_N \cap \partial K} \int_\gamma g\theta_n \, \mathrm{d}s \tag{6.50}$$

whenever $\boldsymbol{x}_n$ is an interior node (case 1) or a Neumann–Neumann node (case 2a). This may be verified as follows. Inserting the definition (6.38) of Δ_K and simplifying reveals that

$$\sum_{K \in \mathcal{P}_n} \widetilde{\Delta}_K(\theta_n) = B(u_X, \theta_n) - (f, \theta_n) - \int_{\Gamma_N} g\theta_n \, \mathrm{d}s$$

and this quantity vanishes thanks to the definition (6.29) of the Galerkin approximation. In those cases where the system (6.46) is singular the least squares solution is selected to avoid numerical difficulties. As a consequence, there exists a constant C, depending only on the number of elements in the patch $\mathcal{P}_n$ such that

$$\sum_{K \in \mathcal{P}_n} \sigma_{K,n}^2 \leq C \sum_{K \in \mathcal{P}_n} \widetilde{\Delta}_K(\theta_n)^2. \tag{6.51}$$

We are now in a position to justify the statements made in Table 6.1. A solution exists in the cases where the topology matrix is singular but the solution is determined only up to the addition of an arbitrary multiple of the vector $\mathbf{1}$. From the point of view of theory, it is irrelevant which particular solution is computed since it is only the *differences* in the components of the solution vector $\boldsymbol{\sigma}$ that are evaluated in the expressions (6.45) for the moments, in those cases where the system (6.46) is indefinite. If the patch intersects the Dirichlet boundary, the topology matrix is nonsingular and the solution is well-defined without any compatibility conditions on the right-hand sides.

6.4.6 Summary

This section has been concerned with developing an efficient and practical numerical procedure for computing boundary fluxes that satisfy the *first-order equilibration condition* (6.32) for Galerkin approximation using first-order elements on a regular partition comprised of quadrilateral or triangular elements. The boundary fluxes are sought in the form of piecewise linear functions on the element edges.

A particularly important step involves selecting the *moments* of the fluxes as the parameters determining the piecewise linear approximations. This leads to the decoupling of the flux computations into independent local problems associated with each of the vertices $\boldsymbol{x}_n$ in the mesh, posed over the small patches $\mathcal{P}_n$ consisting of elements with a vertex located at $\boldsymbol{x}_n$.

While only four distinct cases are possible, it is worthwhile to reformulate the local problems in terms of new unknowns $\{\sigma_{K,n} : K \in \mathcal{P}_n\}$. These are the solution of linear system of equations involving symmetric, positive (in)definite *topology* matrices. Theoretically, the reformulation admits a simple resolution of the problem of nonuniqueness. Practically, the reformulation results in a unified treatment of all the cases and a stable numerical procedure.

The flux moments $\mu^{\gamma}_{K,n}$ are computed from the solution $\{\sigma_{K,n} : K \in \mathcal{P}_n\}$ using the formula (6.45). This is repeated for all vertices in the mesh and the fluxes themselves are recovered either by solving a small algebraic system (6.35) or by using the explicit form (6.36). This results in a set of fluxes $\{g_K\}$ satisfying the first-order equilibration conditions (6.37).

6.5 EFFICIENCY OF THE ESTIMATOR

Theorem 6.1 shows that the estimator obtained by solving local residual problems (6.14) with an equilibrated set of boundary fluxes $\{g_K\}$ will provide an upper bound on the true error. As pointed out earlier, the equilibrated boundary fluxes are not uniquely determined by conditions (6.7) and (6.8), and the quality of the upper bound will depend on the particular selection of boundary fluxes. The purpose of this section is to show that the procedure described in the previous section produces a set of equilibrated boundary fluxes that leads to an estimator that provides two-sided bounds on the true error.

6.5.1 Stability of the Equilibrated Fluxes

The jump discontinuity in the approximation of the normal flux at an interelement boundary is defined by

$$\left[\frac{\partial u_X}{\partial n}\right] = \boldsymbol{n}_K \cdot (\boldsymbol{\nabla} u_X)_K + \boldsymbol{n}_{K'} \cdot (\boldsymbol{\nabla} u_X)_{K'},$$

and the usual interior and boundary residuals r and R are given by

$$r = f + \Delta u_X$$

and

$$R = \begin{cases} -\left[\frac{\partial u_X}{\partial n}\right] & \text{on } \partial K \cap \partial K', \\ 0 & \text{on } \partial K \cap \Gamma_D, \\ g - \frac{\partial u_X}{\partial n} & \text{on } \partial K \cap \Gamma_N. \end{cases}$$

Similarly to Chapter 2, we let $\overline{r}$ and $\overline{R}$ denote the average values of the residuals on the elements and edges respectively.

The next result shows that the procedure described in Section 6.4 is stable in the following sense.

Theorem 6.2 *Suppose that the finite element subspace X is constructed using first-order (linear or bilinear) elements on a regular partitioning $\mathcal{P}$ of the domain Ω into triangular or quadrilateral elements. Let $\{g_K\}$ be the set of equilibrated fluxes satisfying the first-order equilibration conditions, produced by the algorithm described in Section 6.4. Let $\Pi_1^\gamma : L_2(\gamma) \to \mathbb{P}_1(\gamma)$ denote the orthogonal projection with respect to $L_2(\gamma)$. Then, for each element K,*

$$\sum_{\gamma \subset \partial K \setminus \Gamma_N} h_K \left\| g_K - \Pi_1^\gamma \left\langle \frac{\partial u_X}{\partial n_K} \right\rangle \right\|_{L_2(\gamma)}^2$$

$$\leq C \left\{ \|e\|_{\widetilde{K}}^2 + h_K^2 \|r - \overline{r}\|_{L_2(\widetilde{K})}^2 + h_K \sum_{\gamma' \subset \mathrm{int}(\widetilde{K})} \|R - \overline{R}\|_{L_2(\gamma')}^2 \right\} \qquad (6.52)$$

where $\mathrm{int}(\widetilde{K})$ *denotes the interior of the patch $\widetilde{K}$ and the constant C is independent of the mesh size.*

Proof. Let $K \in \mathcal{P}$ be a fixed element. Thanks to the regularity assumption on the partition and Theorem 1.6, the diameter of each element in the patch $\widetilde{K}$ is uniformly equivalent to h_K. Consequently, we need not distinguish between the various mesh sizes and will simply write h for any mesh size throughout the remainder of the argument.

Let $\gamma \subset \partial K$ be any edge of K. Then,

$$\Pi_1^\gamma \left(g_K - \left\langle \frac{\partial u_X}{\partial n_K} \right\rangle \right) \Big|_\gamma \in \mathbb{P}_1(\gamma)$$

and the moments of this quantity satisfy

$$\overset{*}{\mu}{}^\gamma_{K,n} = \int_\gamma \Pi_1^\gamma \left(g_K - \left\langle \frac{\partial u_X}{\partial n_K} \right\rangle \right) \theta_n \, \mathrm{ds} = \int_\gamma \left(g_K - \left\langle \frac{\partial u_X}{\partial n_K} \right\rangle \right) \theta_n \, \mathrm{ds}.$$

By analogy with (6.36),

$$\Pi_1^\gamma \left(g_K - \left\langle \frac{\partial u_X}{\partial n_K} \right\rangle \right) \Big|_\gamma = \overset{*}{\mu}{}^\gamma_{K,\ell} \psi_\ell + \overset{*}{\mu}{}^\gamma_{K,r} \psi_r.$$

Therefore,

$$\left\| \Pi_1^\gamma \left(g_K - \left\langle \frac{\partial u_X}{\partial n_K} \right\rangle \right) \right\|_{L_2(\gamma)} \leq \left| \overset{*}{\mu}{}^\gamma_{K,\ell} \right| \|\psi_\ell\|_{L_2(\gamma)} + \left| \overset{*}{\mu}{}^\gamma_{K,r} \right| \|\psi_r\|_{L_2(\gamma)}$$

and since

$$\|\psi_r\|_{L_2(\gamma)}^2 = \|\psi_\ell\|_{L_2(\gamma)}^2 \leq C h^{-1},$$

it follows that

$$h \left\| \Pi_1^\gamma \left(g_K - \left\langle \frac{\partial u_X}{\partial n_K} \right\rangle \right) \right\|_{L_2(\gamma)}^2 \leq C \sum_{n \in \mathcal{N}(\gamma)} \left| \overset{*}{\mu}{}^\gamma_{K,n} \right|^2. \qquad (6.53)$$

With the aid of (6.49), we conclude that

$$\int_\gamma \left\langle \frac{\partial u_X}{\partial n_K} \right\rangle \theta_n \, \mathrm{d}s = \begin{cases} \frac{1}{2}\left(\tilde{\mu}^\gamma_{K,n} - \tilde{\mu}^\gamma_{K',n}\right) & \text{on } \gamma = \partial K \cap \partial K' \\ \tilde{\mu}^\gamma_{K,n} & \text{on } \gamma = \partial K \cap \Gamma_D \\ \int_\gamma g\theta_n \, \mathrm{d}s & \text{on } \gamma = \partial K \cap \Gamma_N \end{cases}$$

and hence, thanks to (6.45),

$$\overset{*}{\mu}{}^\gamma_{K,n} = \begin{cases} \frac{1}{2}\left(\sigma_{K,n} - \sigma_{K',n}\right) & \text{on } \gamma = \partial K \cap \partial K' \\ \sigma_{K,n} & \text{on } \gamma = \partial K \cap \Gamma_D \\ 0 & \text{on } \gamma = \partial K \cap \Gamma_N \end{cases}$$

where the unknowns $\{\sigma_{K,n}\}$ are determined from conditions (6.46) and satisfy (6.51). It follows that

$$\left|\overset{*}{\mu}{}^\gamma_{K,n}\right|^2 \leq C \sum_{K' \in \mathcal{P}_n} \sigma^2_{K',n} \leq C \sum_{K' \in \mathcal{P}_n} \widetilde{\Delta}_{K'}(\theta_n)^2. \tag{6.54}$$

The terms appearing on the right-hand side may be bounded by first recalling (6.48),

$$\widetilde{\Delta}_{K'}(\theta_n) = B_{K'}(u_X, \theta_n) - (f, \theta_n)_{K'} - \int_{\partial K'} \left\langle \frac{\partial u_X}{\partial n_{K'}} \right\rangle \theta_n \, \mathrm{d}s;$$

then, integrating by parts reveals that

$$\widetilde{\Delta}_{K'}(\theta_n) = -(r, \theta_n)_{K'} - \int_{\partial K'} R\theta_n \, \mathrm{d}s.$$

Applying the Cauchy–Schwarz inequality, it follows that

$$\begin{aligned} \left|\widetilde{\Delta}_{K'}(\theta_n)\right| &\leq \|r\|_{L_2(K')} \|\theta_n\|_{L_2(K')} + \sum_{\gamma' \subset \partial K'} \|R\|_{L_2(\gamma')} \|\theta_n\|_{L_2(\gamma')} \\ &\leq C\left\{ h \|r\|_{L_2(K')} + \sum_{\gamma' \subset \partial K' \cap \mathcal{E}_n} h^{1/2} \|R\|_{L_2(\gamma')} \right\} \end{aligned}$$

since θ_n vanishes on all but those edges $\gamma' \in \mathcal{E}_n$ connected to the node $\boldsymbol{x}_n$ (see definition (6.40)). Hence,

$$\sum_{K' \in \mathcal{P}_n} \left|\widetilde{\Delta}_{K'}(\theta_n)\right|^2 \leq Ch^2 \sum_{K' \in \mathcal{P}_n} \|r\|^2_{L_2(K')} + Ch \sum_{\gamma' \in \mathcal{E}_n} \|R\|^2_{L_2(\gamma')}. \tag{6.55}$$

Therefore, summing over all edges $\gamma \subset \partial K \backslash \Gamma_N$ and combining (6.53), (6.54), and (6.55) leads to the conclusion

$$\begin{aligned} h \sum_{\gamma \subset \partial K} & \left\| g_K - \Pi_1^\gamma \left\langle \frac{\partial u_X}{\partial n_K} \right\rangle \right\|_{L_2(\gamma)}^2 \\ & \leq Ch^2 \sum_{K' \subset \widetilde{K}} \|r\|_{L_2(K')}^2 + Ch \sum_{\gamma' \subset \operatorname{int}(\widetilde{K})} \|R\|_{L_2(\gamma')}^2 \end{aligned}$$

since $g_K|_\gamma \in \mathbb{P}_1(\gamma)$ on all such edges. The result now follows at once thanks to Theorem 2.5. ■

6.5.2 Proof of Efficiency of the Estimator

The stability of the equilibrated fluxes plays a key role in the proof of the efficiency of the estimator.

Theorem 6.3 *Suppose that the partition $\mathcal{P}$ is regular and consists of affine elements. Let $\{g_K\}$ be the set of equilibrated fluxes satisfying the first-order equilibration conditions produced by the algorithm described in Section 6.4, and let $\phi_K \in V_K$ denote the solution of the local residual problem (6.14). Then,*

$$\begin{aligned} \frac{1}{C} \|\phi_K\|_K \leq \|e\|_{\widetilde{K}}^2 & + h_K^2 \|r - \overline{r}\|_{L_2(\widetilde{K})}^2 + h_K \left\|R - \overline{R}\right\|_{L_2(\partial K)}^2 \\ & + h_K \left\| \left\langle \frac{\partial u_X}{\partial n_K} \right\rangle - \Pi_1^\gamma \left\langle \frac{\partial u_X}{\partial n_K} \right\rangle \right\|_{L_2(\partial K \backslash \Gamma_N)}^2 \end{aligned} \tag{6.56}$$

where the constant C is independent of the mesh size.

Proof. Let $v \in V_K$. The data for the local residual problem (6.14) may be rewritten, by integrating by parts, in the form

$$(r, v)_K + \int_{\partial K} \left(g_K - \frac{\partial u_X}{\partial n_K} \right) v \, \mathrm{d}s,$$

or, on rearranging terms, in the equivalent form

$$(r, v)_K + \int_{\partial K} Rv \, \mathrm{d}s + \int_{\partial K} \left(g_K - \left\langle \frac{\partial u_X}{\partial n_K} \right\rangle \right) v \, \mathrm{d}s,$$

and it therefore follows that

$$\begin{aligned} |B_K(\phi_K, v)| & \leq \|r\|_{L_2(K)} \|v\|_{L_2(K)} + \|R\|_{L_2(\partial K)} \|v\|_{L_2(\partial K)} \\ & + \left\| g_K - \left\langle \frac{\partial u_X}{\partial n_K} \right\rangle \right\|_{L_2(\partial K)} \|v\|_{L_2(\partial K)} . \end{aligned} \tag{6.57}$$

Now, by the first-order equilibration condition (6.37), we obtain

$$B_K(\phi_K, 1) = 0,$$

and thus,

$$B_K(\phi_K, v) = B_K(\phi_K, v - \overline{v}) \tag{6.58}$$

where $\overline{v}$ is the (constant) average value of v over element K. Moreover,

$$\|v - \overline{v}\|^2_{L_2(\partial K)} \leq C \|v - \overline{v}\|_{L_2(K)} \|v - \overline{v}\|_{H^1(K)}$$

and with the aid of the Poincaré inequality

$$\|v - \overline{v}\|_{L_2(K)} \leq C h_K |v|_{H^1(K)},$$

it follows that

$$\|v - \overline{v}\|_{L_2(\partial K)} \leq C h_K^{1/2} |v|_{H^1(K)}.$$

These estimates, in conjunction with (6.57) and (6.58), imply that

$$|B_K(\phi_K, v)| \leq C \tag{6.59}$$
$$\left\{ h_K \|r\|_{L_2(K)} + h_K^{1/2} \|R\|_{L_2(\partial K)} + h_K^{1/2} \left\| g_K - \left\langle \frac{\partial u_X}{\partial n_K} \right\rangle \right\|_{L_2(\partial K)} \right\} \| v \|_K.$$

Now, recalling (6.49), it follows that

$$\left\| g_K - \left\langle \frac{\partial u_X}{\partial n_K} \right\rangle \right\|_{L_2(\partial K)} = \left\| g_K - \left\langle \frac{\partial u_X}{\partial n_K} \right\rangle \right\|_{L_2(\partial K \setminus \Gamma_N)}$$

and, by the triangle inequality, this may in turn be bounded by

$$\left\| g_K - \Pi_1^\gamma \left\langle \frac{\partial u_X}{\partial n_K} \right\rangle \right\|_{L_2(\partial K \setminus \Gamma_N)} + \left\| \left\langle \frac{\partial u_X}{\partial n_K} \right\rangle - \Pi_1^\gamma \left\langle \frac{\partial u_X}{\partial n_K} \right\rangle \right\|_{L_2(\partial K \setminus \Gamma_N)}.$$

The proof is completed by inserting this estimate into (6.59), selecting $v = \phi_K$, and applying Theorems 2.5 and 6.2. ∎

Theorem 6.1, together with Theorem 6.3, shows that the estimator provides two-sided bounds on the actual error apart from the familiar, and generally higher-order, terms involving $r - \overline{r}$ and $R - \overline{R}$. The additional terms vanish whenever the element K is the image of the reference element under an *affine* transformation. In general, these terms will not vanish but will usually be of higher-order.

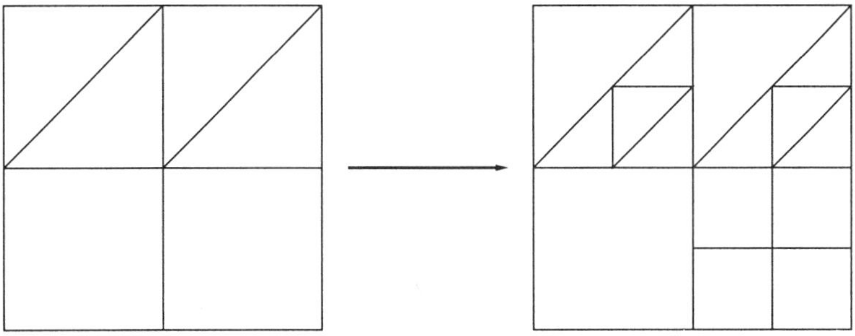

Fig. 6.3 Refinement of triangular and quadrilateral elements into four subelements.

6.6 EQUILIBRATED FLUXES ON PARTITIONS CONTAINING HANGING NODES

Let $\mathcal{P}^0$ be a regular partition consisting of triangular and quadrilateral elements. We now consider a class of partitions containing *hanging nodes*. These are constructed recursively by the following procedure. For $i = 0, 1, \ldots$, a new partition $\mathcal{P}^{i+1}$ is generated by earmarking a subset of the elements in $\mathcal{P}^i$ for refinement. These elements are refined by subdividing into four new elements as shown in Figure 6.3. The nodes in the new partition $\mathcal{P}^{i+1}$ are of two distinct types. A node $\boldsymbol{x}$ is *proper* if it is a vertex of *every* element K such that $\boldsymbol{x} \in \overline{K}$. The remaining nodes are said to be *improper*. Without additional limitations, the final partition $\mathcal{P}$ may contain elements with edges containing arbitrarily many improper nodes. In the interest of simplicity, this possibility is disallowed and the number of improper nodes on any particular edge is limited to at most one. This limitation is realized by refining the elements neighboring element K whenever the refinement of an element K would violate this condition.

The finite element subspace X is constructed on the partition $\mathcal{P}$ based on polynomials of possibly varying degree p. However, appropriate constraints must be applied at the element interfaces in order to maintain continuity [14, 15, 54].

6.6.1 First-Order Equilibration

Suppose that X is based on first-order polynomials. The nodes $\boldsymbol{x}_n$, $n \in \mathcal{N}$, in the partition $\mathcal{P}$ may be partitioned into disjoint sets

$$\mathcal{N}_F = \{n \in \mathcal{N} : \boldsymbol{x}_n \text{ is a proper node}\}$$

and

$$\mathcal{N}_C = \{n \in \mathcal{N} : \boldsymbol{x}_n \text{ is an improper node}\}.$$

The actual degrees of freedom in the finite element subspace X are associated with the set $\mathcal{N}_F$, and identified with function evaluation at the corresponding nodes $\{\boldsymbol{x}_n, n \in \mathcal{N}_F\}$. The value of a function $v \in X$ must be constrained at the nodes $\{\boldsymbol{x}_n, n \in \mathcal{N}_C\}$ in order to maintain continuity across the interface. This means that the finite element approximation satisfies the Galerkin condition at the free, or unconstrained, nodes

$$B(u_X, \theta_n) = (f, \theta_n) + \int_{\Gamma_N} g\theta_n \, \mathrm{d}s \tag{6.60}$$

where θ_n is the Lagrange basis function for X, characterized by the conditions

$$\theta_m(\boldsymbol{x}_n) = \delta_{mn}, \quad m, n \in \mathcal{N}_F. \tag{6.61}$$

However, the corresponding Galerkin condition will generally not be valid for the constrained nodes.

6.6.2 Flux Moments for Unconstrained Nodes

Let $n \in \mathcal{N}_F$ be an unconstrained degree of freedom and let θ_n be the associated Lagrange basis function defined in (6.61). The procedure for the determination of the flux moments with respect to an unconstrained node is similar to the procedure described in Section 6.4, but the presence of improper nodes means that the definitions of the patches (6.39) and edges (6.40) must be modified.

The subdomain influenced by the basis function θ_n is denoted by

$$\Omega_n = \operatorname{int} \operatorname{supp}(\theta_n).$$

This subdomain is composed of a patch of elements

$$\mathcal{P}_n = \{K \in \mathcal{P} : K \subset \Omega_n\}$$

that are interconnected by the element edges contained in the set

$$\mathcal{E}_n = \{\gamma \in \partial\mathcal{P} : \gamma \subset \Omega_n\}.$$

Observe that it is possible for an element K to belong to this set even though $\boldsymbol{x}_n$ might not be a vertex of the element (see Figure 6.4). Similarly, an edge $\gamma \in \mathcal{E}_n$ need not necessarily have an endpoint at the node $\boldsymbol{x}_n$. Nevertheless, if the patch Ω_n contains only proper nodes, then these definitions coincide with the original definitions (6.39) and (6.40).

The subdomain Ω_n may be partitioned into *macroelements*. To this end, we introduce the set $\mathcal{E}_n^*$, defined by

$$\mathcal{E}_n^* = \{\gamma \in \mathcal{E}_n : \gamma \text{ has endpoints at } \boldsymbol{x}_n \text{ and on } \partial\Omega_n\}.$$

It follows that the edges in $\mathcal{E}_n^*$ join the central node $\boldsymbol{x}_n$ to the boundary of the subdomain Ω_n. Consequently, if these edges are removed, then the domain

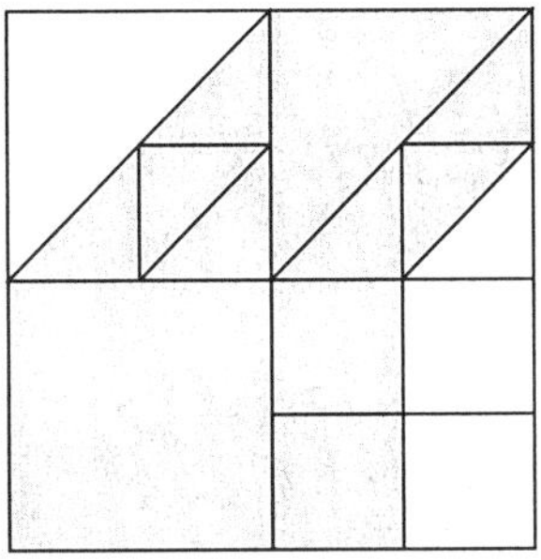

Fig. 6.4 The patch of elements Ω_n forming the support of the basis function associated with central node.

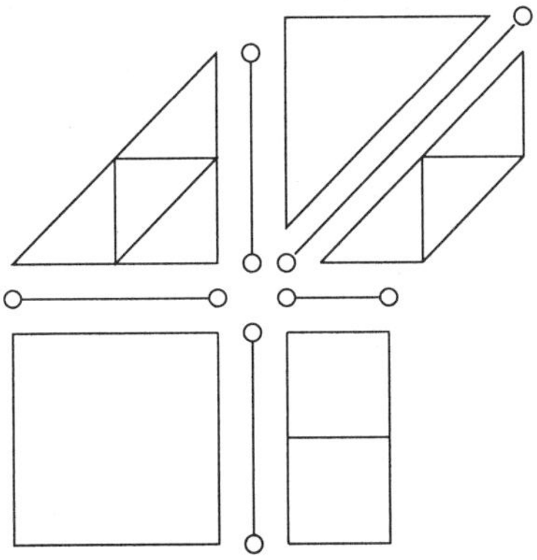

Fig. 6.5 Macroelements $K_n^* \in \mathcal{P}_n^*$ and edges $\gamma \in \mathcal{E}_n^*$ associated with degree of freedom located at the central node.

Ω_n is partitioned into a collection $\mathcal{P}_n^*$ of disjoint subsets $\{K_n^*\}$, which are referred to as the *macroelements associated with* $n \in \mathcal{N}_F$ (see Figure 6.5). More precisely, if

$$\Omega_n^* = \Omega_n - \bigcup_{\gamma \in \mathcal{E}_n^*} \gamma,$$

then Ω_n^* is the union of finitely many macroelements $K_n^* \in \mathcal{P}_n^*$ such that:

- $\overline{\Omega}_n^* = \cup \overline{K}_n^*$.
- If $K_n^* \neq L_n^*$, then $K_n^* \cap L_n^*$ is empty.
- Any nonempty intersection $\partial K_n^* \cap \partial L_n^*$ is an edge $\gamma \in \mathcal{E}_n^*$.
- Each K_n^* either consists of (a) a single element belonging to $\mathcal{P}$ whose vertices are all proper or (b) is the union of a simply connected patch of elements in $\mathcal{P}$.

By analogy with definition (6.38), the residual associated with a macroelement is defined by

$$\Delta_{K_n^*}(\theta_n) = B_{K_n^*}(u_X, \theta_n) - (f, \theta_n)_{K_n^*}.$$

Thanks to the properties of the macroelements, this may be written in an alternative form more convenient for practical computation

$$\Delta_{K_n^*}(\theta_n) = \sum_{K \subset K_n^*} \Delta_K(\theta_n).$$

Consequently, the residuals on the macroelements may be computed by accumulation of residuals over the elements forming the macroelement. Furthermore, observe that if $\boldsymbol{x}_n$ is an interior node (case 1) or is of type Neumann–Neumann (case 2a), then the analogue of the key property (6.50) needed to ensure the existence of equilibrating flux moments is also satisfied:

$$\sum_{K_n^* \in \mathcal{P}_n^*} \Delta_{K_n^*}(\theta_n) = \int_\gamma g\theta_n \, \mathrm{d}s.$$

This is verified by noting that

$$\sum_{K_n^* \in \mathcal{P}_n^*} \Delta_{K_n^*}(\theta_n) = \sum_{K \subset \Omega_n} \Delta_K(\theta_n) = B(u_X, \theta_n) - (f, \theta_n)$$

and the result then follows from the Galerkin property (6.60) as in (6.50).

The data $\mathcal{P}_n^*$, $\mathcal{E}_n^*$, $\Delta_{K_n^*}(\theta_n)$ possess precisely those properties that were needed in Section 6.4 to ensure the existence of a set of flux moments $\mu_{K_n^*,n}^\gamma$ satisfying the first-order equilibration conditions (6.37). Consequently, the same procedures described in Section 6.4 may be invoked in the present situation. The *only* difference is that one works with macroelements rather than with single elements. If the patch contains only proper nodes, then the procedure is identical.

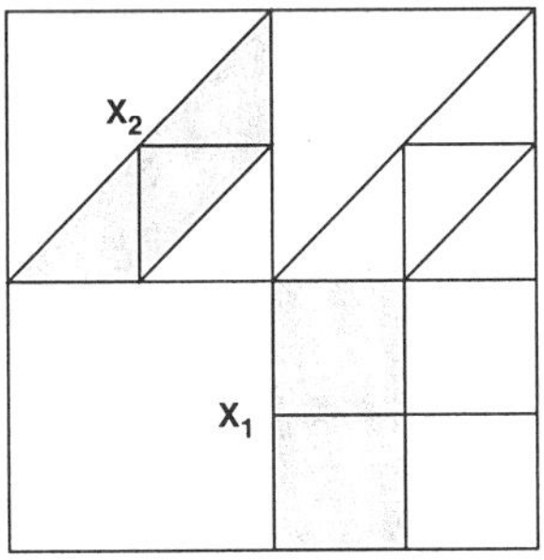

Fig. 6.6 Patches of elements $\mathcal{P}_n$ associated with the constrained nodes shown.

6.6.3 Flux Moments with Respect to Constrained Nodes

Let $n \in \mathcal{N}_C$ be a degree of freedom corresponding to an improper or constrained node. The patch of elements associated with the constrained node is defined to be

$$\mathcal{P}_n = \{K \in \mathcal{P} : n \in \mathcal{N}_K\}.$$

The patch consists of elements of which $\boldsymbol{x}_n$ is a *vertex*, but does not include those elements where $\boldsymbol{x}_n$ lies on an edge (see Figure 6.6). Let $\theta_n^\dagger$ denote the first-order Lagrange function associated with the vertices of the elements in the patch $\mathcal{P}_n$; that is,

$$\theta_m^\dagger(\boldsymbol{x}_n) = \delta_{mn} \quad \boldsymbol{x}_m, \boldsymbol{x}_n \in \cup_{\overline{K} \in \mathcal{P}_n} K.$$

The superscript † reflects the fact that these functions do not belong to the finite element space X (since they are discontinuous at the constrained interface). These functions are related to the element basis functions before the connectivity constraints have been applied.

The first-order equilibration conditions for elements in this patch lead to a set of conditions on the flux moments that is identical in form to those for a Dirichlet–Dirichlet vertex (case 2c). Observe that no additional conditions are needed on the residuals to guarantee the existence of the flux moments in this case. Indeed, the compatibility condition (6.60) is not valid in this situation. Consequently, the procedures of Section 6.4 may be invoked to produce a set of flux moments with respect to the functions $\theta_n^\dagger$, satisfying the conditions (6.37).

6.6.4 Recovery of Actual Fluxes

The procedures described above result in sets of flux moments $\tilde{\mu}^\gamma_{K_n^*,n}$ associated with the basis function θ_n over the macroelements K_n^*, and flux moments with respect to the functions $\theta_n^\dagger$ over the constrained elements K. How are these moments used to reconstruct the actual flux functions?

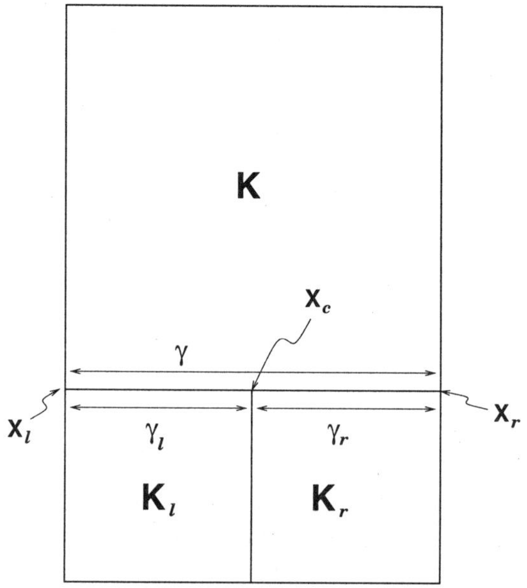

Fig. 6.7 Configuration of elements and edges used to illustrate the reconstruction of the equilibrated fluxes on an edge γ containing an improper node.

If no improper nodes are located on an edge $\gamma \subset \partial K$ with endpoints at $\boldsymbol{x}_\ell$ and $\boldsymbol{x}_r$, then both moments of the flux g_K with respect to the basis functions θ_ℓ and θ_r are available. The procedure described in Section 6.4.2 may be applied to recover the actual flux g_K on the edge.

The more interesting situation where the edge γ contains a single improper node $\boldsymbol{x}_c$ is shown in Figure 6.7. The following data are available:

- The flux moments with respect to the basis functions θ_ℓ and θ_r over the entire edge γ are known from the treatment of the patches associated with nodes $\boldsymbol{x}_\ell$ and $\boldsymbol{x}_r$, that is, $\mu^{\gamma}_{K,\ell}$ and $\mu^{\gamma}_{K,r}$.

- The flux moments with respect to the function $\theta^{\dagger}_{c}$ over the portions γ_ℓ and γ_r are known from the treatment of the patch associated with the constrained node $\boldsymbol{x}_c$, that is, $\mu^{\gamma_\ell}_{K_\ell,\ell}$ and $\mu^{\gamma_r}_{K_r,r}$.

The unknown element fluxes g_K, g_{K_ℓ}, and g_{K_r} satisfy

$$g_K = \begin{cases} -g_{K_\ell} & \text{on } \gamma_\ell \\ -g_{K_r} & \text{on } \gamma_r, \end{cases} \tag{6.62}$$

and it therefore suffices to reconstruct g_{K_ℓ} and g_{K_r}. This relation shows that the flux g_K will be a discontinuous piecewise linear function on the edge γ. By symmetry, it suffices to apply the same reconstruction procedure for the flux

g_{K_ℓ}. The flux moment of g_{K_ℓ} with respect to the node $\boldsymbol{x}_c$ is already available:

$$\int_{\gamma_\ell} g_{K_\ell}\theta_c^\dagger \, \mathrm{d}s = \mu_{K_\ell,\ell}^{\gamma_\ell}.$$

Accordingly, we need only obtain the flux moment with respect to the node $\boldsymbol{x}_\ell$ to be in a position to apply the procedure of Section 6.4.2.

This moment is obtained by firstly noting that on the interface γ we obtain

$$\theta_\ell = \begin{cases} \theta_\ell^\dagger + \frac{1}{2}\theta_c^\dagger & \text{on } \gamma_\ell \\ \frac{1}{2}\theta_c^\dagger & \text{on } \gamma_r \end{cases}$$

where $\theta_\ell^\dagger$ is the elemental function associated with the nodes $\boldsymbol{x}_\ell$ and $\boldsymbol{x}_c$ of element K_ℓ. Therefore,

$$\mu_{K,\ell}^{\gamma} = \int_\gamma g_K \theta_\ell \, \mathrm{d}s = \int_{\gamma_\ell} g_K \left(\theta_\ell^\dagger + \frac{1}{2}\theta_c^\dagger\right) \mathrm{d}s + \int_{\gamma_r} g_K \frac{1}{2}\theta_c^\dagger \, \mathrm{d}s.$$

With the aid of (6.62), this may be rewritten as

$$-\int_{\gamma_\ell} g_{K_\ell} \left(\theta_\ell^\dagger + \frac{1}{2}\theta_c^\dagger\right) \mathrm{d}s - \int_{\gamma_r} g_{K_r} \frac{1}{2}\theta_c^\dagger \, \mathrm{d}s$$

and then, by definition, it follows that

$$\mu_{K,\ell}^{\gamma} = -\mu_{K_\ell,\ell}^{\gamma_\ell} - \frac{1}{2}\mu_{K_\ell,c}^{\gamma_\ell} - \frac{1}{2}\mu_{K_r,c}^{\gamma_r}.$$

Hence, the desired flux moment is given in terms of the known data by

$$\mu_{K_\ell,\ell}^{\gamma_\ell} = -\mu_{K,\ell}^{\gamma} - \frac{1}{2}\mu_{K_\ell,c}^{\gamma_\ell} - \frac{1}{2}\mu_{K_r,c}^{\gamma_r}.$$

The remaining flux moment is given by

$$\mu_{K_r,r}^{\gamma_r} = -\mu_{K,r}^{\gamma} - \frac{1}{2}\mu_{K_\ell,c}^{\gamma_\ell} - \frac{1}{2}\mu_{K_r,c}^{\gamma_r}.$$

The procedure described in Section 6.4.2 may then be applied to obtain g_{K_ℓ} and g_{K_r}, and hence g_K.

6.7 EQUILIBRATED FLUXES FOR HIGHER-ORDER ELEMENTS

The discussion has thus far been limited to finite element approximation involving first-order elements. However, in the first-order equilibration procedure presented in Section 6.4, the only property that was really required of the finite element approximation u_X was the satisfaction of the Galerkin condition (6.29) with respect to the first-order functions. A Galerkin approximation based on higher-order elements also possesses this property. It follows that precisely the same procedure described in Section 6.4 could be employed to recover fluxes satisfying the first-order equilibration conditions (6.32) from a given finite element approximation based on any order elements.

The definition of the *first-order* equilibration conditions (6.32) may be generalized to *p th-order equilibration conditions* as follows:

$$\left.\begin{aligned}(f,\theta_n)_K - B_K(u_X,\theta_n) + \int_{\partial K} g_K\theta_n\,\mathrm{d}s &= 0 \quad \forall n \in \mathcal{N}(K)\\ g_K + g_{K'} &= 0 \text{ on } \partial K \cap \partial K'\\ g_K &= g \text{ on } \partial K \cap \Gamma_N.\end{aligned}\right\} \tag{6.63}$$

While this appears to be identical to (6.32), it is important to note that the set of degrees of freedom $\mathcal{N}(K)$ appearing in the statement now includes interior and edge nodes in addition to vertices. Conditions (6.63) are therefore much stricter than (6.32).

For a pth-order finite element approximation, the Galerkin property holds with respect to all the basis functions up to and including the full order p. However, this additional property was superfluous in the construction of boundary fluxes $\{g_K\}$ that fulfill the first-order equilibration condition using the procedure described in Section 6.4. This prompts the following question: *Is it possible to exploit the full Galerkin property and thereby reconstruct boundary fluxes $\{g_K\}$ that fulfill the p th-order equilibration conditions?* The purpose of this section is to show that this is indeed the case and also to show how the previous algorithm may be extended to achieve this goal.

Let $\mathcal{V}(\gamma)$ denote the degrees of freedom associated with the nodes located at the endpoints of an edge γ and let $\mathcal{E}(\gamma)$ denote the degrees of freedom associated with the nodes located on the interior of the edge. It will be useful to recall some elementary facts concerning the basis functions associated with the nodes located on the interior of an edge γ. Firstly, the only basis functions θ_n that are not identically zero an edge γ are the vertex functions ($n \in \mathcal{V}(\gamma)$) and edge functions ($n \in \mathcal{E}(\gamma)$). Each edge function θ_n, $n \in \mathcal{E}(\gamma)$, vanishes at the endpoints of the edge γ and on all remaining edges in the partition. In particular, if γ is an edge of element K and $n \in \mathcal{E}(\gamma)$, then

$$\theta_n = \begin{cases} \text{polynomial of degree at most } p \text{ on } \gamma \text{ which} \\ \text{vanishes on the remaining edges } \partial K\backslash\gamma. \end{cases} \tag{6.64}$$

Finally, the vertex and edge functions on γ together form a basis for the space $\mathbb{P}_p(\gamma)$ consisting of polynomials of degree at most p on the edge,

$$\mathbb{P}_p(\gamma) = \operatorname{span}\{\theta_n|_\gamma, n \in \mathcal{V}(\gamma) \cup \mathcal{E}(\gamma)\}. \tag{6.65}$$

6.7.1 The Form of the Boundary Fluxes

The final condition of (6.63) dictates that the boundary flux on an edge $\gamma \subset \Gamma_N$ should coincide with the Neumann data on the edge. By analogy with the form adopted for the boundary fluxes in Section 6.4 for first-order equilibration, the boundary fluxes used to attain pth-order equilibration are chosen to be piecewise polynomials of degree p on the remaining edges. In view of the property (6.65), we may write

$$g_K|_\gamma \begin{cases} \text{is the Neumann data } g \text{ on } \gamma \subset \Gamma_N, \text{ or} \\ \text{belongs to } \operatorname{span}\{\theta_n|_\gamma, n \in \mathcal{V}(\gamma) \cup \mathcal{E}(\gamma)\} \text{ on remaining edges.} \end{cases}$$

In order to characterize a polynomial of degree p, it is necessary to specify $p+1$ independent degrees of freedom. As before, it is convenient to select the degrees of freedom to be moments of the flux weighted against the basis functions on the edge:

$$\mu_{K,n}^\gamma = \int_\gamma g_K \theta_n \, \mathrm{d}s, \quad n \in \mathcal{V}(\gamma) \cup \mathcal{E}(\gamma). \tag{6.66}$$

The actual flux itself is easily determined from these moments by writing

$$g_K|_\gamma = \sum_{n \in \mathcal{V}(\gamma) \cup \mathcal{E}(\gamma)} \alpha_n \theta_n$$

where $\{\alpha_n\}$ are constants. The values of these constants are determined by definition (6.66) through the conditions

$$\sum_{n \in \mathcal{V}(\gamma) \cup \mathcal{E}(\gamma)} (\theta_m, \theta_n)_\gamma \, \alpha_n = \mu_{K,m}^\gamma \quad \forall m \in \mathcal{V}(\gamma) \cup \mathcal{E}(\gamma) \tag{6.67}$$

where $(\cdot,\cdot)_\gamma$ denote the inner product on $L_2(\gamma)$. As before, this may be formulated as a matrix equation involving the *edge mass matrix* $\boldsymbol{M}_\gamma$ whose entries are defined by

$$\boldsymbol{M}_\gamma = [(\theta_m, \theta_n)_\gamma].$$

The matrix is nonsingular thanks to the linear independence of the basis functions on the edge, and so the coefficients $\{\alpha_n\}$ are uniquely determined by the conditions (6.67). For the practical implementation, it is often more efficient to precompute the inverse $\boldsymbol{M}_\gamma$ for the range of polynomial degrees of interest.

6.7.2 Determination of the Flux Moments

The selection of the degrees of freedom in the boundary fluxes means that we must determine the flux moments $\mu_{K,n}^\gamma$ defined by

$$\mu_{K,n}^\gamma = \int_\gamma g_K \theta_n \, \mathrm{d}s, \quad n \in \mathcal{V}(\gamma) \cap \mathcal{E}(\gamma).$$

We now make a key observation. A careful examination of the algorithm described in Section 6.4 shows that the precise form of the vertex basis functions was not necessary for the computation of the *moments*. As a matter of fact, the particular form of the basis functions only enters when the actual flux function g_K is reconstructed following the steps described in Section 6.4.2. As a consequence, *the flux moments with respect to the vertex functions θ_n, $n \in \mathcal{V}(\gamma)$, may be determined using the same algorithm described in Section 6.4.*

Thanks to this observation, it only remains to determine the flux moments with respect to the edge functions θ_n, $n \in \mathcal{E}(\gamma)$. The pth-order equilibration conditions impose the following constraints on the flux moments associated with this function:

$$\left.\begin{aligned} \mu^{\gamma}_{K,n} &= \Delta_K(\theta_n) & \forall K : \gamma \subset \partial K \\ \mu^{\gamma}_{K,n} + \mu^{\gamma}_{K',n} &= 0 & \gamma = \partial K \cap \partial K' \\ \mu^{\gamma}_{K,n} &= \textstyle\int_\gamma g\theta_n \,\mathrm{d}s & \gamma = \partial K \cap \Gamma_N \end{aligned}\right\} \tag{6.68}$$

where Δ_K is defined as in (6.38).

The conditions (6.68) are similar in spirit to the conditions (6.37) determining the moments with respect to the vertex functions. However, the patches associated with the element edges are much simpler than those associated with the element vertices, and they consist of at most two elements. As was the case for vertices, the conditions (6.68) overdetermine the flux moments and the existence of a solution hinges on the compatibility of the data. There are three cases to be considered depending on the location of the edge γ.

1. *Interior Edge:* If $\gamma = \partial K \cap \partial K'$, then conditions (6.68) require that

$$\left.\begin{aligned} \mu^{\gamma}_{K,n} &= \Delta_K(\theta_n) \\ \mu^{\gamma}_{K',n} &= \Delta_{K'}(\theta_n) \\ \mu^{\gamma}_{K,n} + \mu^{\gamma}_{K',n} &= 0. \end{aligned}\right\}$$

The first two equations determine the flux moments uniquely so it suffices to check that the final condition holds. This is verified by noting that θ_n is supported on the pair of elements K and K' so that with the aid of (6.38),

$$\Delta_K(\theta_n) + \Delta_{K'}(\theta_n) = B(u_X, \theta_n) - (f, \theta_n)$$

and this quantity vanishes thanks to the Galerkin property (6.29).

2. *Neumann Edge:* If $\gamma \subset \partial K \cap \Gamma_N$, then conditions (6.68) require that

$$\left.\begin{aligned} \mu^{\gamma}_{K,n} &= \Delta_K(\theta_n) \\ \mu^{\gamma}_{K,n} &= \textstyle\int_\gamma g\theta_n \,\mathrm{d}s. \end{aligned}\right\}$$

These equations are compatible provided that the following quantity vanishes

$$\Delta_K(\theta_n) - \int_\gamma g\theta_n \, \mathrm{d}s = B_K(u_X, \theta_n) - (f, \theta_n)_K - \int_\gamma g\theta_n \, \mathrm{d}s.$$

This follows at once from the Galerkin property (6.29) after noting that, in this case, θ_n is supported on the single element K.

3. Dirichlet Edge: If $\gamma \subset \Gamma_D$, then conditions (6.68) reduce to the single equation

$$\mu_{K,n}^\gamma = \Delta_K(\theta_n)$$

since no additional conditions are applied to the flux on a Dirichlet boundary.

Thus, in all cases, there exists a unique solution of the conditions (6.68) that is determined directly in terms of the local residuals $\Delta_K(\theta_n)$. Consequently, it follows that boundary fluxes may be determined such that the pth-order equilibration conditions are satisfied. The formal proof that the fluxes verify the conditions follows from the above arguments and the results of Section 6.4.

In conclusion, these developments show that it is indeed possible to recover boundary fluxes from a pth-order finite element approximation that satisfy the pth-order equilibration conditions. The essential differences with the first-order case are that the boundary fluxes are sought in the form of piecewise polynomials of degree p on the edges, with the degrees of freedom again chosen to be the moments with respect to the basis functions. The moments associated with the vertices are obtained using precisely the same algorithm described in Section 6.4, while the moments associated with the higher-order edge functions are obtained directly from the element residuals Δ_K by the formula

$$\mu_{K,n}^\gamma = \Delta_K(\theta_n) - (f, \theta_n)_K \quad \forall n \in \mathcal{E}(\gamma). \tag{6.69}$$

The fluxes $\{g_K\}$ themselves are recovered from the moments by solving the linear system (6.67).

6.8 BIBLIOGRAPHICAL REMARKS

The notion of post-processing the finite element approximation to obtain equilibrating fluxes or tractions on the element boundaries goes back to the work of Ladevèze and Leguillon [77], Kelly [73], and Bank and Weiser [37]. Ladevèze and Leguillon [77] proposed *a posteriori* error estimators by solving a local dual problem in conjunction with the equilibrated fluxes as boundary conditions. Bank and Weiser [37] obtained estimators for affine approximation on linear triangular elements, by solving the primal problem and, on the basis of numerical results, conjectured that the resulting estimator is always an upper bound on the actual error (Theorem 6.1 in the text). This conjecture was proved in Ainsworth and Oden [11] in the general setting of *hp*-finite element approximation. The proof given in the text and the proof of stability of the equilibrated flux procedure is based on Ainsworth and Babuška [5].

Equilibration procedures for elliptic systems were discussed in [8, 9, 10]. The treatment of the equilibration conditions on meshes with hanging nodes is based on reference [7].

7

Methodology for the Comparison of Estimators

7.1 INTRODUCTION

Faced with such a plethora of *a posteriori* error estimators, the practitioner might reasonably wonder which estimator to select for a particular application. Consequently, it is desirable to have a methodology for the comparison of estimators. Equally well, when a new *a posteriori* error estimator has been proposed, the traditional approach has been to feature the effectivity indices obtained when the estimator has been applied to a particular problem where the true error is known. Benchmark computations can provide valuable information. However, the benchmarks must be carefully selected to identify all traits, both desirable and otherwise, of the error estimator under consideration. It is possible to construct examples where a particular estimator is superior when applied with a certain set of data on a particular partition, but whose performance is markedly inferior when the estimator is applied to a wider set of data and meshes.

One is led to the conclusion that it is essential to have an objective and standardized means to assess the quality of an estimator that exercises all the features of the particular estimator. The purpose of the present chapter is to give an overview of the steps that have been made in this direction. The theory underlying the ideas is presented in a simplified form with the purpose of motivating the methodology. The computational procedure is presented in detail with the aim of providing sufficient information for the user to implement and apply the approach to their own area of application.

It should be borne in mind that the methodology to be presented has its limitations. In particular, the procedure allows the evaluation of the extreme effectivity indices for the estimator when certain effects such as the influence of singularities, the effects due to the boundary of the domain and mesh grading have been isolated. Moreover, the effectivity indices are those that would be obtained in the asymptotic limit when the mesh size tends to zero. The preasymptotic behavior of the estimators might well lead to rather different conclusions concerning the suitability of a particular estimator. One may view the role of the systematic testing to be presented *not* as a means to justify an estimator, but rather as a minimal criterion the estimator must meet if it is not to be dismissed out of hand.

Subject to these limitations, the approach to be described goes some way to avoiding the pitfalls of making general statements on the performance of an estimator based on the performance for a limited number of test problems.

7.2 OVERVIEW OF THE TECHNIQUE

This section is concerned with providing a general description of the approach before embarking on a detailed account of the theoretical foundations. As usual, the discussion will be focused on the model problem of Section 1.4:

$$-\Delta u = f \text{ in } \Omega; \quad u = 0 \text{ on } \partial\Omega.$$

The solution is approximated in the usual fashion with the finite element approximation u_X sought in the subspace $X \subset H_0^1(\Omega)$ constructed on a regular partition of the domain, such that

$$u_X \in X : \quad B(u_X, v_X) = (f, v_X) \quad \forall v_X \in X.$$

For the purposes of exposition, it will be assumed throughout this chapter that the finite element subspace is constructed using piecewise affine functions on triangular elements. However, generalization to other types of element and spaces is possible.

An accurate finite element resolution of this problem generally requires the partition to be sufficiently refined in the neighborhood of the boundary in order to resolve the singularities due to the geometry and changes in type of boundary condition. However, if the data f are smooth, then the mesh on the interior of the domain can be essentially uniform. An example of the type of mesh we have in mind is shown in Figure 7.1.

The aim of the analysis is to focus on the asymptotic behavior of a given *a posteriori* error estimator on convex interior patches $\Omega_2 \subset \Omega_1$ over which the mesh is regular. (In the sequel, a further stronger assumption of *translation invariance* will be made on the mesh in the interior subdomain.) To facilitate a local analysis over an interior subdomain, it will be necessary to make assumptions concerning the properties of the true solution u over the subdomain Ω_0:

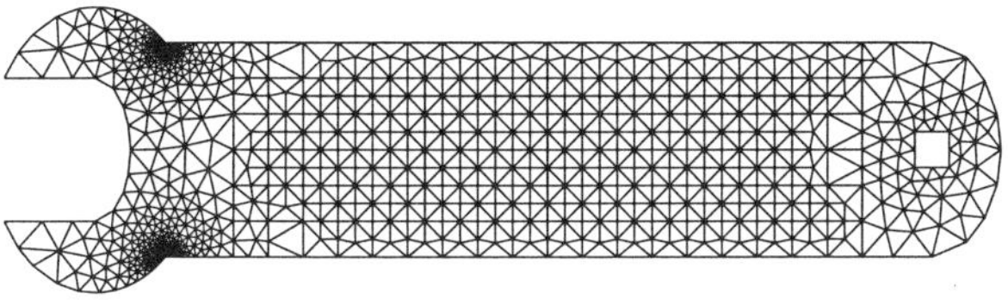

Fig. 7.1 Typical example of a mesh that is composed of translation invariant cells on interior subdomains yet is locally refined in the neighborhood of re-entrant corners to control the pollution effects of the singularities.

($A1$) The true solution u satisfies a regularity condition on the interior subdomain, $u \in W^{3,\infty}(\Omega_0)$, and there exists a positive constant μ_0 such that

$$\inf_{\boldsymbol{x}\in\Omega_0} \sum_{|\beta|=2} \left|D^\beta u(\boldsymbol{x})\right|^2 \geq \mu_0^2 > 0. \tag{7.1}$$

This condition rules out trivial cases when the second derivatives of the solution vanish identically. In such cases, a first-order finite element approximation would perform as if a quadratic approximation were used.

As has already been alluded to, in order to isolate the behavior of the finite element approximation on the interior subdomain from the influence of outside effects such as singularities in the solution around re-entrant corners, the mesh must be sufficiently refined in the neighborhood of such features [114]. This requirement is embodied in the following assumption concerning the rate of convergence of the approximation measured in the L_2-norm over the subdomain:

($A2$) There exists $\tau \in (3/2, 2)$ such that

$$\|u - u_X\|_{L_2(\Omega_1)} \leq Ch^\tau D_1 \tag{7.2}$$

where C is a positive constant that depends on u but is independent of the diameter D_1 of the domain Ω_1 and the mesh size h.

This assumption ensures that the accuracy of the approximation over the subdomain is not adversely affected by *pollution* effects due to outside influences [114], yet the assumption does not preclude the possibility of the solution u having singularities outside the subdomain. Generally speaking, in the complete absence of pollution effects, condition (7.2) holds for all values of τ up to 2.

The smoothness condition on the true solution means that u may be approximated well by a quadratic polynomial $q \in \mathbb{Q}_2$ over the subdomain Ω_1. In particular, it will be assumed that q is the polynomial obtained by expanding

u as a Taylor series about the centroid of Ω_1, so that

$$\begin{aligned} |u-q|_{W^{1,\infty}(\Omega_1)} &\leq CD_1^2 \|u\|_{W^{3,\infty}(\Omega_1)} \\ |u-q|_{W^{2,\infty}(\Omega_1)} &\leq CD_1 \|u\|_{W^{3,\infty}(\Omega_1)} \end{aligned} \tag{7.3}$$

for some constant C. A local finite element approximation may be developed for q over the subdomain Ω_1 by seeking the solution $q_X^* \in X(\Omega_1)$ of the problem

$$\left.\begin{aligned} B_{\Omega_1}(q-q_X^*, v) &= 0 \quad \forall v \in X(\Omega_1) \\ \textstyle\int_{\Omega_1}(q-q_X^*)dx &= 0. \end{aligned}\right\} \tag{7.4}$$

where

$$B_{\Omega_1}(u,v) = \int_{\Omega_1} \nabla u \cdot \nabla v \, dx. \tag{7.5}$$

In other words, q_X^* is the orthogonal projection of q onto the space $X(\Omega_1)$ with respect to the energy inner product. The fact that no essential boundary conditions are imposed means that the second condition in (7.4) is necessary to ensure that q_X^* is unique by fixing the average value. The space consisting of the restrictions of functions belonging to the original finite element subspace X to a subdomain $\Omega' \subset \Omega$ is denoted by $X(\Omega')$,

$$X(\Omega') = \{v|_{\Omega'} : v \in X\}$$

while the space consisting of functions with support in Ω' is denoted by $X_0(\Omega')$

$$X_0(\Omega') = \{v \in X : \operatorname{supp}(v) \subset \Omega'\}.$$

Let $I_X v$ denote the piecewise affine interpolant to a continuous function v at the element vertices.

Under appropriate assumptions on the subdomains Ω_1 and Ω_2, it will later be shown that there exists $\sigma \in (0, 1/5)$, depending on the exponent τ in assumption $(A2)$, such that

$$\left| \|e\|_{\Omega_2} - \|q - q_X^*\|_{\Omega_2} \right| \leq Ch^\sigma \|e\|_{\Omega_2}. \tag{7.6}$$

The significance of this result is that the principal contribution to the error e over the subdomain Ω_2 arises from the approximation error inherent in the Galerkin projection itself. That is to say, the error arising in the approximation $q \approx q_X^*$ cancels out the leading-order term in the error e in the actual Galerkin approximation over the subdomain (since the relative difference converges to zero as $h \to 0$).

At this stage we are prepared to present a heuristic description of the key idea in this chapter. Let us suppose we have an *a posteriori* error estimator that acts on the data f and the finite element approximation u_X to produce an estimator $\epsilon(u_X, f, \Omega_2)$ for the error $\|e\|_{\Omega_2}$ over the subdomain Ω_2. Equally well, the estimator may be applied to the quadratic solution q and its approximation q_X^* to produce an estimator $\epsilon(q_X^*, -\Delta q, \Omega_2)$ for the error $\|q - q_X^*\|_{\Omega_2}$. Now, prompted by the estimate (7.6), one is led to conjecture that, in the limit $h \to 0$, the effectivity indices of each of these estimates will be the *same*, that is,

$$\lim_{h\to 0} \frac{\epsilon(u_X, f, \Omega_2)}{\|e\|_{\Omega_2}} = \lim_{h\to 0} \frac{\epsilon(q_X^*, -\Delta q, \Omega_2)}{\|q - q_X^*\|_{\Omega_2}}. \tag{7.7}$$

Later it will be shown that this intuitive notion is essentially correct, provided that the estimator satisfies certain mild conditions.

The practical significance of this result is that one may compute the extreme values of the effectivity index

$$\lim_{h\to 0} \frac{\epsilon(q_X^*, -\Delta q, \Omega_2)}{\|q - q_X^*\|_{\Omega_2}} \tag{7.8}$$

over all possible quadratic polynomials q and the class of all meshes that will be used in practical computation. This would automatically mean that the asymptotic values of the effectivity indices for every possible infinite-dimensional solution u were included in the range of values obtained by computing effectivity indices in the case where the solution belongs to the finite-dimensional space consisting of quadratic polynomials. The range of asymptotic effectivity indices for various a posteriori error estimators would provide an objective comparison of the estimators.

The procedure, as it has been described above, contains one major drawback: One would need to have access to the solution q_X^* in the limiting case when $h \to 0$. Obviously, this is not a practical proposition. Therefore, an approach will be developed using *translation invariant* meshes, whereby one need only compute the solution of a local finite element problem formulated over a single reference cell. The solution on the reference cell is then used to construct the *asymptotic finite element* approximation q_X^{asy} to q for which it will be shown that, under analogous conditions to those needed to obtain (7.6),

$$\left| \|e\|_{\Omega_2} - \|q - q_X^{\text{asy}}\|_{\Omega_2} \right| \leq Ch^\sigma \|e\|_{\Omega_2}. \tag{7.9}$$

Naturally, conclusions identical to the ones drawn from (7.6) concerning q_X^* may be drawn for q_X^{asy} on the basis of (7.9). The difference is that q_X^{asy} is practically computable.

7.3 APPROXIMATION OVER AN INTERIOR SUBDOMAIN

7.3.1 Translation Invariant Meshes

Let $\Omega_0 \subset \Omega$ denote an interior subdomain over which the mesh consists of a repeating pattern of basic *mesh cells* of size h as shown, for example, in

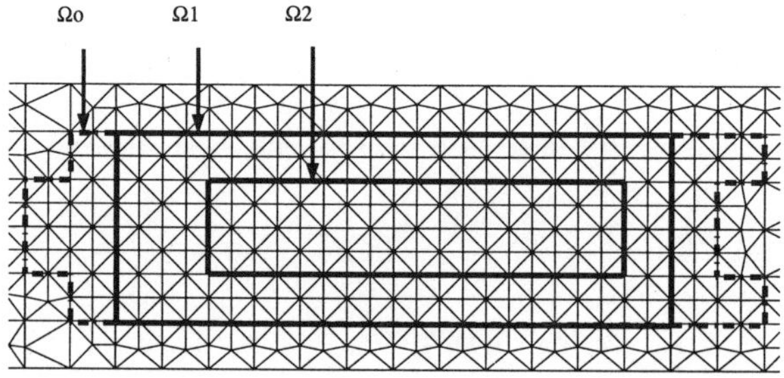

Fig. 7.2 Interior subdomains $\Omega_2 \subset \Omega_1 \subset \Omega_0$ composed of translation invariant cells. The inner subdomains Ω_1 and Ω_2 are assumed to be convex while Ω_0 may be nonconvex.

Figure 7.2. The centroids of the cells are denoted by $\boldsymbol{c}_\alpha$, $\alpha \in \mathcal{A}_0$, where $\mathcal{A}_0$ is a suitable indexing set, and the cell centered at $\boldsymbol{c}_\alpha$ is denoted by $\mathcal{C}(\boldsymbol{c}_\alpha, h)$. In particular, the cells are disjoint

$$\mathcal{C}(\boldsymbol{c}_\alpha, h) \cap \mathcal{C}(\boldsymbol{c}_\beta, h) = \emptyset \quad \text{for } \alpha \neq \beta$$

and form a partition of Ω_0,

$$\Omega_0 = \bigcup_{\alpha \in \mathcal{A}_0} \overline{\mathcal{C}(\boldsymbol{c}_\alpha, h)}. \tag{7.10}$$

Furthermore, suppose that subdomains $\Omega_2 \subset \Omega_1 \subset \Omega_0 \subset \Omega$ and indexing sets $\mathcal{A}_2 \subset \mathcal{A}_1 \subset \mathcal{A}_0$ are chosen so that

$$\Omega_k = \bigcup_{\alpha \in \mathcal{A}_k} \overline{\mathcal{C}(\boldsymbol{c}_\alpha, h)}, \quad k \in \{1, 2\},$$

where Ω_1 and Ω_2 are convex and regular in the sense that the largest ball that may be inscribed inside Ω_k has diameter at least CD_k, where D_k is the diameter of Ω_k and C is a positive constant.

By applying a translation of $-\boldsymbol{c}_\alpha$ followed by a scaling of $1/h$, the particular cell $\mathcal{C}(\boldsymbol{c}_\alpha, h)$ is transformed into a *reference cell* $\mathcal{C}(\boldsymbol{0}, 1)$ centered at the origin (see Figure 7.3). The preimages, under this transformation, of the elements contained in the cell $\mathcal{C}(\boldsymbol{c}_\alpha, h)$ form a partitioning of the reference cell, denoted by $\mathcal{P}$.

The mesh on the subdomain Ω_0 is said to be *translation invariant* if the meshes on all of the cells $\mathcal{C}(\boldsymbol{c}_\alpha, h)$, $\alpha \in \mathcal{A}_0$, are the image of the *same* mesh $\mathcal{P}$ on the same basic reference cell $\mathcal{C}(\boldsymbol{0}, 1)$. In particular, this means that the nodes of the partition on opposite edges of the cells lie in the same positions relative to the endpoints of the edge. For instance, cells of the type shown in Figure 7.4 are inadmissible, since the mesh obtained by piecing together

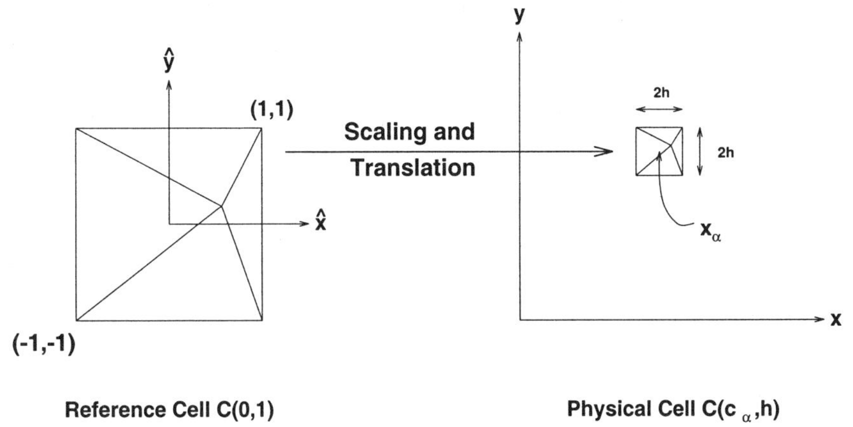

Fig. 7.3 Each of the physical cells $C(\boldsymbol{c}_\alpha, h)$ is the image of a single reference cell under a scaling of h followed by a translation of $\boldsymbol{c}_\alpha$.

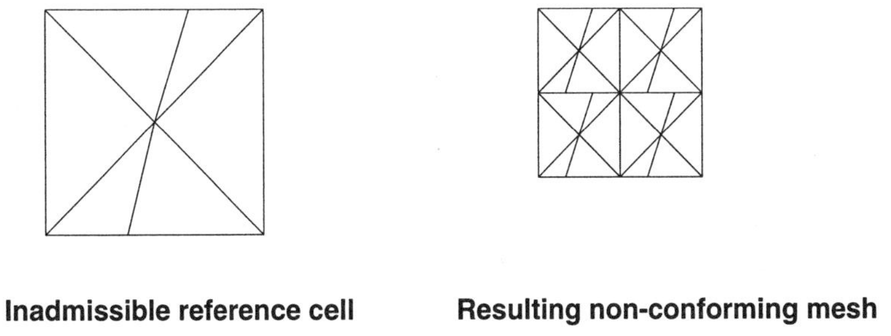

Fig. 7.4 Example of an inadmissible mesh cell that would result in a nonconforming mesh. For an admissible cell, the relative positions of the nodes on opposite edges must match.

such cells would result in a nonconforming finite element partition on the subdomain Ω_0.

The finite element subspace associated with the partition $\mathcal{P}$ of the reference cell is denoted by $\widehat{X}$. For simplicity, it will be assumed that the basic reference cell $\mathcal{C}(\mathbf{0}, 1)$ is a square of side 2, so that the physical cells are squares of side $2h$. The theory could readily be extended to allow hexahedral or rhomboidal cells, or indeed any convex domain that tessellates the plane.

7.3.2 Lower Bounds on the Error

The purpose of this section is to establish lower bounds on the rate of convergence of the finite element approximation.

Lemma 7.1 *Let u satisfy assumption (A1) and let e denote the error in the finite element approximation u_X. Then, for $h \leq h_0$ sufficiently small, there exist positive constants C depending on μ_0, h_0, and $||u||_{W^{3,\infty}(\Omega_0)}$, such that for all $\alpha \in \mathcal{A}_0$,*

$$|||e|||_{\mathcal{C}(\mathbf{c}_\alpha, h)} \geq Ch^2 \tag{7.11}$$

and

$$||e||_{L_2(\mathcal{C}(\mathbf{c}_\alpha, h))} \geq Ch^3. \tag{7.12}$$

These results are a special case of the following more general result originally given in reference [19].

Lemma 7.2 *Let u satisfy condition (A1). Then, for $h \leq h_0$ sufficiently small, there exists a constant $C > 0$ depending on μ_0, h_0, and $||u||_{W^{3,\infty}(\Omega_0)}$, such that for all $\alpha \in \mathcal{A}_0$,*

$$\inf_{v \in X(\mathcal{C}(\mathbf{c}_\alpha, h))} |||u - v|||_{\mathcal{C}(\mathbf{c}_\alpha, h)} \geq Ch^2 \tag{7.13}$$

and

$$\inf_{v \in X(\mathcal{C}(\mathbf{c}_\alpha, h))} ||u - v||_{L_2(\mathcal{C}(\mathbf{c}_\alpha, h))} \geq Ch^3. \tag{7.14}$$

Proof. Let K be any element contained in the cell $\mathcal{C}(\mathbf{c}_\alpha, h)$. Expanding u as a Taylor series about the centroid $\overline{\boldsymbol{x}}$ of the element gives

$$u(\boldsymbol{x}) = u_0(\boldsymbol{x}) + \frac{1}{2} \sum_{|\beta|=2} D^\beta u(\overline{\boldsymbol{x}})(\boldsymbol{x} - \overline{\boldsymbol{x}})^\beta + r(\boldsymbol{x})$$

where $u_0 \in \mathbb{P}_1(K)$ and the remainder satisfies

$$|r(\boldsymbol{x})| \leq C|\boldsymbol{x} - \overline{\boldsymbol{x}}|^3$$

and

$$\max_{|\beta|=1} |D^\beta r(\boldsymbol{x})| \leq C|\boldsymbol{x} - \overline{\boldsymbol{x}}|^2$$

where C is a positive constant depending on $\|u\|_{W^{3,\infty}(K)}$ but not on the size h of the element.

Let the quadratic polynomial $\overline{u}$ be defined by

$$\overline{u}(\boldsymbol{x}) = u_0(\boldsymbol{x}) + \frac{1}{2}\sum_{|\beta|=2} D^\beta u(\overline{\boldsymbol{x}})(\boldsymbol{x}-\overline{\boldsymbol{x}})^\beta.$$

Now, we assert that

$$\inf_{v\in\mathbb{P}_1(K)} \|\overline{u}-v\|_K \geq Ch\,|\overline{u}|_{H^2(K)}$$

where the constant C is independent of h. This may be proved in the case $h=1$ by first observing that the quantities appearing on each side of the inequality define norms over the space $\mathbb{Q}_2\backslash\mathbb{P}_1$. The result in the case $h\neq 1$ then follows with the aid of a scaling argument (similarly to the proof of Corollary 1.2).

Furthermore, since (A1) implies that

$$|\overline{u}|^2_{H^2(K)} \geq Ch^2 \inf_{\boldsymbol{x}\in\Omega_0} \sum_{|\beta|=2} \left|D^\beta u(\boldsymbol{x})\right|^2 \geq Ch^2\mu_0^2$$

it follows that

$$\inf_{v\in\mathbb{P}_1(K)} \|\overline{u}-v\|_K \geq Ch^2\mu_0.$$

Hence, by the triangle inequality,

$$\inf_{v\in\mathbb{P}_1(K)} \|u-v\|_K \geq \inf_{v\in\mathbb{P}_1(K)} \|\overline{u}-v\|_K - \|r\|_K \geq Ch^2\mu_0 - C^*h^3.$$

Obviously,

$$\inf_{v\in X(\mathcal{C}(\boldsymbol{c}_\alpha,h))} \|u-v\|^2_{\mathcal{C}(\boldsymbol{c}_\alpha,h)} \geq \sum_{K\subset\mathcal{C}(\boldsymbol{c}_\alpha,h)} \inf_{v\in\mathbb{P}_1(K)} \|u-v\|^2_K \geq \left(Ch^2\mu_0 - C^*h^3\right)^2$$

where the constants depend on the (fixed) number of elements in each cell. Therefore, there exists $h_0>0$ such that if $h\leq h_0$, then

$$\inf_{v\in X(\mathcal{C}(\boldsymbol{c}_\alpha,h))} \|u-v\|_{\mathcal{C}(\boldsymbol{c}_\alpha,h)} \geq Ch^2.$$

The proof in the case of the L_2-norm proceeds along the same lines. ∎

7.3.3 Interior Estimates

The main tool in deriving a convergence estimate for the difference between the true error e and $q-q^*_X$ over the subdomain Ω_2 will be the technique of interior estimates as described, for example, in the survey article of Wahlbin [114], where the following result will be found.

Lemma 7.3 *Let $\Omega_2 \subset \Omega_1 \subset \Omega$ satisfy*

$$d = \text{dist}(\partial\Omega_1, \partial\Omega_2) \geq Ch \tag{7.15}$$

for some positive constant C. Let $\psi \in H^1(\Omega_1)$ satisfy the condition

$$B(\psi, v) = 0 \quad \forall v \in X_0(\Omega_1). \tag{7.16}$$

Then,

$$\|\psi\|_{\Omega_2} \leq C \inf_{v \in X(\Omega_1)} \left\{ \|\psi - v\|_{\Omega_1} + \frac{1}{d} \|\psi - v\|_{L_2(\Omega_1)} \right\} + \frac{C}{d} \|\psi\|_{L_2(\Omega_1)} \tag{7.17}$$

where C is a positive constant that is independent of h and d.

Proof. See Corollary 9.1 in reference [114]. ∎

Firstly, we derive bounds for the error e over an interior subdomain.

Theorem 7.4 *Suppose conditions ($A1$) and ($A2$) hold, let $\sigma = \frac{2}{5}(\tau - \frac{3}{2})$ and let q_X^* be defined as in (7.4). If the diameters D_1 and D_2 of the subdomains $\Omega_2 \subset \Omega_1 \subset \Omega$ satisfy*

$$ch^{(1+\sigma)/2} \leq D_2 \leq D_1 \leq Ch^{(1+\sigma)/2} \tag{7.18}$$

and the distance between the subdomains satisfies

$$d = \text{dist}(\partial\Omega_1, \partial\Omega_2) \geq Ch^\sigma D_1 \tag{7.19}$$

for positive constants c and C independent of h, then

$$ch^{(3+\sigma)/2} \leq \|e\|_{\Omega_2} \leq Ch^{(3+\sigma)/2} \tag{7.20}$$

where c and C are positive constants depending on $\|u\|_{W^{3,\infty}(\Omega_1)}$ and μ_0.

Proof. Applying Lemma 7.1, we obtain

$$\|e\|^2_{\Omega_2} = \sum_{\alpha \in \mathcal{A}_2} \|e\|^2_{C(c_\alpha, h)} \geq C \sum_{\alpha \in \mathcal{A}_2} h^4$$

and then the lower bound in (7.20) follows since

$$\sum_{\alpha \in \mathcal{A}_2} h^2 \geq C|\Omega_2| \geq CD_2^2 \geq Ch^{1+\sigma}.$$

The proof of the upper bound consists of preparing the ground for an application of Lemma 7.3. Observe that distance d between subdomain boundaries satisfies $d \geq Ch^\sigma D_1 \geq Ch^{(1+3\sigma)/2} \geq Ch$ as $h \to 0$, since $\sigma \leq 1/5$. Then, thanks to $d \geq Ch^{(1+3\sigma)/2}$, assumption ($A2$) implies that

$$\begin{aligned} \frac{1}{d} \|u - u_X\|_{L_2(\Omega_1)} &\leq Ch^{\tau - (1+3\sigma)/2} D_1 \\ &\leq Ch^{\tau - (1+3\sigma)/2 + (1+\sigma)/2} \end{aligned}$$

and substituting for $\tau = (5\sigma + 3)/2$ gives

$$\frac{1}{d} \|u - u_X\|_{L_2(\Omega_1)} \leq Ch^{3(1+\sigma)/2}. \tag{7.21}$$

Applying the standard approximation properties of the interpolant gives

$$|\!|\!| u - I_X u |\!|\!|_{\Omega_1} \leq Ch \|u\|_{H^2(\Omega_1)} \leq ChD_1 \leq Ch^{(3+\sigma)/2}.$$

Similarly,

$$\frac{1}{d} \|u - I_X u\|_{L_2(\Omega_1)} \leq \frac{Ch}{d} h^{(3+\sigma)/2}$$

and hence, since $d \geq Ch$,

$$\frac{1}{d} \|u - I_X u\|_{L_2(\Omega_1)} \leq Ch^{(3+\sigma)/2}.$$

The Galerkin orthogonality property of the error implies, *a fortiori*, that

$$B_{\Omega_1}(e, v) = 0 \quad \forall v \in X_0(\Omega_1);$$

thus on applying Lemma 7.3 and selecting $v = I_X u - u_X \in X(\Omega_1)$, we conclude that

$$\begin{aligned} |\!|\!| e |\!|\!|_{\Omega_2} \leq\ & C |\!|\!| u - I_X u |\!|\!|_{\Omega_1} \\ & + \frac{C}{d} \|u - I_X u\|_{L_2(\Omega_1)} \\ & + \frac{C}{d} \|e\|_{L_2(\Omega_1)} \end{aligned}$$

thanks to (7.21). Inserting the bounds derived above then gives,

$$|\!|\!| e |\!|\!|_{\Omega_2} \leq Ch^{(3+\sigma)/2}$$

as required. ■

The key result of this section is as follows:

Theorem 7.5 *Suppose conditions (A1) and (A2) hold, let* $\sigma = \frac{2}{5}(\tau - \frac{3}{2})$ *and let* q_X^* *be defined as in (7.4). If the diameters* D_1 *and* D_2 *of the subdomains* $\Omega_2 \subset \Omega_1 \subset \Omega$ *satisfy*

$$ch^{(1+\sigma)/2} \leq D_2 \leq D_1 \leq Ch^{(1+\sigma)/2} \tag{7.22}$$

and the distance between the subdomains satisfies

$$d = \operatorname{dist}(\partial\Omega_1, \partial\Omega_2) \geq Ch^{\sigma} D_1 \tag{7.23}$$

for positive constants c *and* C *independent of* h*, then*

$$\left| |\!|\!| e |\!|\!|_{\Omega_2} - |\!|\!| q - q_X^* |\!|\!|_{\Omega_2} \right| \leq |\!|\!| e - (q - q_X^*) |\!|\!|_{\Omega_2} \leq Ch^{\sigma} |\!|\!| e |\!|\!|_{\Omega_2} \tag{7.24}$$

where C is a positive constant depending on $\|u\|_{W^{3,\infty}(\Omega_1)}$ and μ_0.

Proof. Once again, the proof consists of preparing the ground for an application of Lemma 7.3. It follows at once from (7.21) and (7.20) that

$$\frac{1}{d}\|e\|_{L_2(\Omega_1)} \leq Ch^{3(1+\sigma)/2-(3+\sigma)/2} \|\!|e|\!\|_{\Omega_2} \leq Ch^{\sigma} \|\!|e|\!\|_{\Omega_2}. \tag{7.25}$$

Applying the Aubin–Nitsche duality argument [50] on the convex subdomain Ω_1 leads to the conclusion

$$\|q - q_X^*\|_{L_2(\Omega_1)} \leq Ch \|\!|q - q_X^*|\!\|_{\Omega_1}.$$

Then, by the usual optimality property of the Galerkin approximation,

$$\|\!|q - q_X^*|\!\|_{\Omega_1} \leq \|\!|q - I_X q|\!\|_{\Omega_1}$$

where $I_X q$ is the interpolant to q on Ω_1. Furthermore, by the standard approximation properties of the interpolant,

$$\|\!|q - I_X q|\!\|_{\Omega_1} \leq Ch\|q\|_{H^2(\Omega_1)} \leq ChD_1 \leq Ch^{(3+\sigma)/2}.$$

Consequently, since $d \geq Ch^{(1+3\sigma)/2}$, with the aid of (7.20) we obtain

$$\frac{1}{d}\|q - q_X^*\|_{L_2(\Omega_1)} \leq Ch^{1+(3+\sigma)/2-(1+3\sigma)/2-(3+\sigma)/2} \|\!|e|\!\|_{\Omega_2} = Ch^{(1-3\sigma)/2} \|\!|e|\!\|_{\Omega_2}$$

and then, since $\sigma \leq 1/5$, it follows that $(1-3\sigma)/2 \geq 1/5 \geq \sigma$ and so

$$\frac{1}{d}\|q - q_X^*\|_{L_2(\Omega_1)} \leq Ch^{\sigma} \|\!|e|\!\|_{\Omega_2}. \tag{7.26}$$

The main result is obtained as follows. By definition (7.4) of q_X^* and properties of the Galerkin approximation u_X, it easily follows that

$$B_{\Omega_1}(u - q - (u_X - q_X^*), v) = 0 \quad \forall v \in X_0(\Omega_1);$$

thus on applying Lemma 7.3 and selecting $v = I_X(u-q) \in X(\Omega_1)$, we conclude that

$$\begin{aligned} \|\!|u - q - (u_X - q_X^*)|\!\|_{\Omega_2} \leq\; & C\|\!|u - q - I_X(u-q)|\!\|_{\Omega_1} \\ & + \frac{C}{d}\|u - q - I_X(u-q)\|_{L_2(\Omega_1)} \\ & + \frac{C}{d}\|u - q - (u_X - q_X^*)\|_{L_2(\Omega_1)}. \end{aligned} \tag{7.27}$$

The first term is estimated using properties of the interpolant and recalling (7.3)

$$\|\!|u - q - I_X(u-q)|\!\|_{\Omega_1} \leq Ch\|u - q\|_{H^2(\Omega_1)}$$

$$
\begin{aligned}
&\leq ChD_1 \|u - q\|_{W^{2,\infty}(\Omega_1)} \\
&\leq ChD_1^2 \|u\|_{W^{3,\infty}(\Omega_1)} \\
&\leq Ch^{1+(1+\sigma)-(3+\sigma)/2} \| e \|_{\Omega_2} \\
&\leq Ch^{(1+\sigma)/2} \| e \|_{\Omega_2} \\
&\leq Ch^{\sigma} \| e \|_{\Omega_2}
\end{aligned}
$$

since $(1+\sigma)/2 \geq 1/2 > \sigma$. The second term is treated in a similar fashion:

$$
\begin{aligned}
\frac{1}{d} \|u - q - I_X(u-q)\|_{L_2(\Omega_1)} &\leq Ch^{2-(1+3\sigma)/2} \|u - q\|_{H^2(\Omega_1)} \\
&\leq Ch^{2-(1+3\sigma)/2} D_1 \|u - q\|_{W^{2,\infty}(\Omega_1)} \\
&\leq Ch^{2-(1+3\sigma)/2} D_1^2 \|u\|_{W^{3,\infty}(\Omega_1)} \\
&\leq Ch^{2-(1+3\sigma)/2+(1+\sigma)-(3+\sigma)/2} \| e \|_{\Omega_2} \\
&\leq Ch^{1-\sigma} \| e \|_{\Omega_2} \\
&\leq Ch^{\sigma} \| e \|_{\Omega_2}
\end{aligned}
$$

since $1 - \sigma \geq \sigma$. The final term is estimated by the triangle inequality,

$$
\|u - q - (u_X - q_X^*)\|_{L_2(\Omega_1)} \leq \|u - u_X\|_{L_2(\Omega_1)} + \|q - q_X^*\|_{L_2(\Omega_1)}
$$

and then applying the bounds (7.25) and (7.26). Inserting these estimates into (7.27) shows that

$$
\| u - q - (u_X - q_X^*) \|_{\Omega_2} \leq Ch^{\sigma} \| e \|_{\Omega_2}
$$

and then, by the triangle inequality, we have

$$
\left| \| e \|_{\Omega_2} - \| q - q_X^* \|_{\Omega_2} \right| \leq \| u - q - (u_X - q_X^*) \|_{\Omega_2},
$$

and the result follows as claimed. ■

7.4 ASYMPTOTIC FINITE ELEMENT APPROXIMATION

This section is concerned with the development of an approximation q_X^{asy} to q that has the same approximation properties as the finite element approximation q_X^* over the subdomain Ω_2 but, in addition, is computable in the asymptotic case $h \to 0$.

7.4.1 Periodic Finite Element Projection on Reference Cell

The space of periodic functions on the reference cell $\mathcal{C}(\mathbf{0}, 1)$ is defined by

$$
H^{1,\mathrm{per}}(\mathcal{C}(\mathbf{0},1)) = \left\{ v \in H^1(\mathcal{C}(\mathbf{0},1)) : \begin{array}{l} v(x,-1) = v(x,+1), \\ v(-1,y) = v(+1,y), \end{array} x, y \in (-1,1) \right\} \tag{7.28}
$$

and is equipped with the usual norm for the space $H^1(\mathcal{C}(\mathbf{0},1))$. The corresponding finite element subspace constructed using the partition $\mathcal{P}$ of the reference cell is defined by

$$\widehat{X}^{\text{per}} = \left\{v \in H^{1,\text{per}}(\mathcal{C}(\mathbf{0},1)) : v|_K \in \mathbb{P}_1(K) \quad \forall K \in \mathcal{P}\right\}$$

and consists of piecewise affine functions on the reference mesh.

The periodic finite element projection

$$\widehat{\Pi}^{\text{per}} : H^{1,\text{per}}(\mathcal{C}(\mathbf{0},1)) \to \widehat{X}^{\text{per}}$$

is defined for each $\widehat{u} \in H^{1,\text{per}}(\mathcal{C}(\mathbf{0},1))$ to be the unique function $\widehat{z} \in \widehat{X}^{\text{per}}$ satisfying the conditions

$$\left.\begin{aligned} \widehat{B}(\widehat{u}-\widehat{z},\widehat{v}) &= 0 \quad \forall \widehat{v} \in \widehat{X}^{\text{per}} \\ \int_{\mathcal{C}(\mathbf{0},1)} (\widehat{u}-\widehat{z})d\widehat{\boldsymbol{x}} &= 0 \end{aligned}\right\} \tag{7.29}$$

where $\widehat{B}(\cdot,\cdot)$ denotes the bilinear form evaluated over the reference cell. The function $\widehat{z}$ is well-defined since the bilinear form is continuous and coercive on the (closed) subspace of $\widehat{X}^{\text{per}}$ consisting of functions whose average value vanishes.

7.4.2 Periodic Finite Element Projection on a Physical Cell

The periodic finite element projection associated with a particular physical mesh cell $\mathcal{C}(\boldsymbol{c}_\alpha, h)$, $\alpha \in \mathcal{A}_0$, will now be considered. Firstly, the periodic Sobolev space $H^{1,\text{per}}(\mathcal{C}(\boldsymbol{c}_\alpha,h))$ associated with the mesh cell may be characterized by

$$H^{1,\text{per}}(\mathcal{C}(\boldsymbol{c}_\alpha,h)) = \left\{v : v(\boldsymbol{c}_\alpha + h\bullet) \in H^{1,\text{per}}(\mathcal{C}(\mathbf{0},1))\right\}.$$

Here, the notation $v(\boldsymbol{c}_\alpha + h\bullet)$ indicates the function defined by the rule $\boldsymbol{x} \to v(\boldsymbol{c}_\alpha + h\boldsymbol{x})$, where $\boldsymbol{x}$ is a dummy variable. Evidently, the space $H^{1,\text{per}}(\mathcal{C}(\boldsymbol{c}_\alpha,h))$ consists of functions that are periodic on the physical cell $\mathcal{C}(\boldsymbol{c}_\alpha,h)$. Similarly, the finite element subspace associated with the cell is defined by

$$X_\alpha^{\text{per}} = \left\{v : v(\boldsymbol{c}_\alpha + h\bullet) \in \widehat{X}^{\text{per}}\right\}.$$

The periodic finite element projection

$$\Pi_\alpha^{\text{per}} : H^{1,\text{per}}(\mathcal{C}(\boldsymbol{c}_\alpha,h)) \to X_\alpha^{\text{per}}$$

of a function $u \in H^{1,\text{per}}(\mathcal{C}(\boldsymbol{c}_\alpha,h))$ is defined to be the unique function $z_\alpha \in X_\alpha^{\text{per}}$ satisfying the conditions

$$\left.\begin{aligned} B_{\mathcal{C}(\boldsymbol{c}_\alpha,h)}(u-z_\alpha,v) &= 0 \quad \forall v \in X_\alpha^{\text{per}} \\ \int_{\mathcal{C}(\boldsymbol{c}_\alpha,h)} (u-z_\alpha)\,\mathrm{d}\boldsymbol{x} &= 0 \end{aligned}\right\} \tag{7.30}$$

and, as before, the function z_α is well-defined.

Let $u \in H^{1,\text{per}}(\mathcal{C}(\boldsymbol{c}_\alpha, h))$, then we may define $\widehat{u} \in H^{1,\text{per}}(\mathcal{C}(\mathbf{0}, 1))$ by the rule

$$\widehat{u}(\widehat{\boldsymbol{x}}) = u(\boldsymbol{c}_\alpha + h\widehat{\boldsymbol{x}}), \quad \widehat{\boldsymbol{x}} \in \mathcal{C}(\mathbf{0}, 1).$$

A simple change of variable reveals that the periodic finite element projection of u on the cell $\mathcal{C}(\boldsymbol{c}_\alpha, h)$ and the projection of the function $\widehat{u}$ on the reference cell are related by the rule

$$\left(\widehat{\Pi}^{\text{per}}\widehat{u}\right)(\widehat{\boldsymbol{x}}) = \left(\Pi_\alpha^{\text{per}} u\right)(\boldsymbol{c}_\alpha + h\widehat{\boldsymbol{x}}), \quad \widehat{\boldsymbol{x}} \in \mathcal{C}(\mathbf{0}, 1).$$

7.4.3 Periodic Extension on a Subdomain

Let $\Omega_0 \subset \Omega$ be the subdomain composed of mesh cells $\mathcal{C}(\boldsymbol{c}_\alpha, h)$, $\alpha \in \mathcal{A}_0$, as in (7.10). Let $\widehat{u} \in H^{1,\text{per}}(\mathcal{C}(\mathbf{0}, 1))$ be any given periodic function over the reference cell. A periodic function $u_\alpha \in H^{1,\text{per}}(\mathcal{C}(\boldsymbol{c}_\alpha, h))$ may be defined over each of the mesh cells in Ω, by the rule

$$u_\alpha(\boldsymbol{x}) = \widehat{u}(h^{-1}(\boldsymbol{x} - \boldsymbol{c}_\alpha)), \quad \boldsymbol{x} \in \mathcal{C}(\boldsymbol{c}_\alpha, h).$$

The interface conditions (7.28) characterizing the space $H^{1,\text{per}}(\mathcal{C}(\mathbf{0}, 1))$ mean that the functions u_α match on the cell interfaces. Therefore, these functions may be pieced together to produce $\widetilde{u} \in H^1(\Omega_0)$ defined by

$$\widetilde{u}(\boldsymbol{x}) = u_\alpha(\boldsymbol{x}), \quad \boldsymbol{x} \in \mathcal{C}(\boldsymbol{c}_\alpha, h)).$$

The function $\widetilde{u}$ is called the *periodic extension* associated with $\widehat{u}$ on the subdomain Ω_0. The space of all periodic extensions over Ω_0 is denoted by $\widetilde{H}^{1,\text{per}}(\Omega_0)$, and is defined by

$$\widetilde{H}^{1,\text{per}}(\Omega_0) = \left\{ v \in H^1(\Omega_0) : v|_{\mathcal{C}(\boldsymbol{c}_\alpha, h))} = \widehat{v}(h^{-1}(\bullet - \boldsymbol{c}_\alpha)) \text{ with } \widehat{v} \in H^{1,\text{per}}(\mathcal{C}(\mathbf{0}, 1)) \right\}.$$

The corresponding finite element subspace of $\widetilde{H}^{1,\text{per}}(\Omega_0)$ is obtained by constructing periodic extensions of functions belonging to the periodic finite element subspace $\widehat{X}^{\text{per}}$,

$$\widetilde{X}^{\text{per}} = \left\{ v \in \widehat{X}^{\text{per}} : v|_{\mathcal{C}(\boldsymbol{c}_\alpha, h))} = \widehat{v}(h^{-1}(\bullet - \boldsymbol{c}_\alpha)) \quad \text{for some } \widehat{v} \in \widehat{X}^{\text{per}} \right\}.$$

A projection operator associated with these spaces,

$$\widetilde{\Pi}^{\text{per}} : \widetilde{H}^{1,\text{per}}(\Omega_0) \to \widetilde{X}^{\text{per}},$$

is defined so that if $\widetilde{u} \in \widetilde{H}^{1,\text{per}}(\Omega_0)$, then $\widetilde{\Pi}^{\text{per}}\widetilde{u} = \widetilde{z}$, where $\widetilde{z} \in \widetilde{X}^{\text{per}}$ satisfies the conditions

$$\left.\begin{aligned} B_{\Omega_0}(\widetilde{u} - \widetilde{z}, \widetilde{v}) &= 0 \quad \forall \widetilde{v} \in \widetilde{X}^{\text{per}} \\ \int_{\Omega_0} (\widetilde{u} - \widetilde{z})\, \mathrm{d}\boldsymbol{x} &= 0. \end{aligned}\right\} \tag{7.31}$$

Once again, $\widetilde{z}$ is well-defined by these conditions.

The correspondence between the space of periodic extensions and the space of periodic functions on the reference cell means that the projection $\widetilde{\Pi}^{\mathrm{per}}$ is related to the periodic finite element projection on the reference cell as is demonstrated by the following result:

Lemma 7.6 *Let $\widetilde{u} \in \widetilde{H}^{1,\mathrm{per}}(\Omega_0)$ be the periodic extension associated with the function $\widehat{u} \in H^{1,\mathrm{per}}(\mathcal{C}(\mathbf{0},1))$. Let z be defined by*

$$z|_{\mathcal{C}(\boldsymbol{c}_\alpha,h)} = \left(\widehat{\Pi}^{\mathrm{per}}\widehat{u}\right)(h^{-1}(\bullet - \boldsymbol{c}_\alpha)).$$

Then $z \in \widetilde{X}^{\mathrm{per}}$ and

$$\left.\begin{aligned} B_{\Omega_0}(\widetilde{u} - z, \widetilde{v}) &= 0 \quad \textit{for } \widetilde{v} \in \widetilde{X}^{\mathrm{per}} \\ \textstyle\int_{\Omega_0} (\widetilde{u} - z)\,\mathrm{d}\boldsymbol{x} &= 0. \end{aligned}\right\} \tag{7.32}$$

Consequently, $z = \widetilde{\Pi}^{\mathrm{per}} u$.

Proof. The first assertion follows at once from the definition of the space $\widetilde{X}^{\mathrm{per}}$ and the fact that $\widehat{\Pi}^{\mathrm{per}}\widehat{u}$ belongs to $\widehat{X}^{\mathrm{per}}$.

Let $\widetilde{v} \in \widetilde{X}^{\mathrm{per}}$. Then, by definition, there exists $\widehat{v} \in \widehat{X}^{\mathrm{per}}$ such that $\widetilde{v}$ is the periodic extension of $\widehat{v}$. Now, by direct manipulation,

$$\begin{aligned} B_{\Omega_0}(\widetilde{u} - z, \widetilde{v}) &= \sum_{\alpha \in \mathcal{A}_0} B_{\mathcal{C}(\boldsymbol{c}_\alpha,h)}(\widetilde{u} - z, \widetilde{v}) \\ &= \sum_{\alpha \in \mathcal{A}_0} \widehat{B}(\widehat{u} - \widehat{\Pi}^{\mathrm{per}}\widehat{u}, \widehat{v}) \\ &= 0 \end{aligned}$$

where the first property (7.29) of the projection $\widehat{\Pi}^{\mathrm{per}}$ has been invoked.

The final assertion follows in a similar fashion by decomposing the integral over Ω_0 into integrals over the individual cells, applying a change variable to obtain integrals over the reference cell and applying the second property (7.29) of the projection $\widehat{\Pi}^{\mathrm{per}}$. ∎

This result shows that the computation of the periodic finite element projection of a function $\widetilde{u} \in \widetilde{H}^{1,\mathrm{per}}(\Omega_0)$ may be reduced to the problem of determining the periodic finite element projection over the reference cell. This is the crucial observation that will allow the ideas presented in Section 7.2 to be brought to fruition.

7.4.4 Asymptotic Finite Element Approximation

The quadratic polynomial approximation to u is of the form

$$q(\boldsymbol{x}) = \sum_{|\beta| \leq 2} a_\beta \boldsymbol{x}^\beta = \sum_{|\beta| \leq 2} a_\beta x^{\beta_1} y^{\beta_2}$$

where $\beta = (\beta_1, \beta_2)$ is a multiindex and $|\beta| = \beta_1 + \beta_2$. The following result shows that the difference $q - I_X q$ is actually a periodic extension over Ω_0.

Lemma 7.7 *Let q be a quadratic polynomial over Ω_0 of the form*

$$q(\boldsymbol{x}) = \sum_{|\beta| \le 2} a_\beta \boldsymbol{x}^\beta, \tag{7.33}$$

then $q - I_X q \in \widetilde{H}^{1,\mathrm{per}}(\Omega_0)$.

Proof. According to the definition of the space $\widetilde{H}^{1,\mathrm{per}}(\Omega_0)$, it suffices to show the existence of a function $\widehat{r} \in H^{1,\mathrm{per}}(\mathcal{C}(\boldsymbol{0}, 1))$ such that for each cell $\mathcal{C}(\boldsymbol{c}_\alpha, h)$,

$$(q - I_X q)|_{\mathcal{C}(\boldsymbol{c}_\alpha, h)} = \widehat{r}(h^{-1}(\boldsymbol{x} - \boldsymbol{c}_\alpha)).$$

Let $\widehat{\mathcal{I}}\widehat{\boldsymbol{x}}^\beta$ denote the piecewise affine interpolant of the function $\widehat{\boldsymbol{x}}^\beta$ on the partition $\mathcal{P}$ of the reference cell. The difference

$$\delta_\beta(\widehat{\boldsymbol{x}}) = \widehat{\boldsymbol{x}}^\beta - \widehat{\mathcal{I}}\widehat{\boldsymbol{x}}^\beta$$

vanishes identically except in the cases where $|\beta| = 2$. Consider the case where $\widehat{\boldsymbol{x}}^\beta = \widehat{x}^2$. This function is constant on the left- and right-hand edges of the reference cell and so δ_β vanishes on each of these faces. On the top and bottom edges of the cell, δ_β vanishes at each of the mesh points on the edges and has the same constant second derivative between the nodes along each of the edges between the nodes. Therefore, since the relative positions of the mesh points on the top and bottom edges match, it follows that

$$\delta_\beta(\widehat{x}, -1) = \delta_\beta(\widehat{x}, +1), \quad \widehat{x} \in (-1, 1).$$

Similar arguments may be applied to the remaining case when $|\beta| = 2$, and we therefore conclude $\delta_\beta \in \widehat{X}^{\mathrm{per}}$ for all $|\beta| \le 2$. Consequently, we may select $\widehat{r} \in \widehat{X}^{\mathrm{per}}$ to be the function

$$\widehat{r}(\widehat{\boldsymbol{x}}) = h^2 \sum_{|\beta|=2} a_\beta (\widehat{\boldsymbol{x}}^\beta - \widehat{\mathcal{I}}\widehat{\boldsymbol{x}}^\beta).$$

Let $\boldsymbol{x} \in \mathcal{C}(\boldsymbol{c}_\alpha, h)$ and denote

$$d_\alpha(\boldsymbol{x}) = (q - I_X q)(\boldsymbol{x}) - \widehat{r}(h^{-1}(\boldsymbol{x} - \boldsymbol{c}_\alpha)).$$

The quadratic terms in d_α cancel and therefore d_α is piecewise affine over the cell $\mathcal{C}(\boldsymbol{c}_\alpha, h)$. Furthermore, both of the terms in d_α vanish separately when the point $\boldsymbol{x}$ coincides with a vertex of the mesh in the cell. Consequently, it follows that d_α vanishes identically over the cell, and the result follows as claimed. ∎

The fact that the difference $q - I_X q$ is a periodic extension whenever q is a quadratic polynomial means that the quantity

$$q_X^{\mathrm{asy}} = I_X q + \widetilde{\Pi}^{\mathrm{per}}(q - I_X q) \tag{7.34}$$

is well-defined. This function is called the *asymptotic finite element approximation* of q over the subdomain Ω_0. Basic properties of the asymptotic finite element approximation are recorded in the following:

Lemma 7.8 *Let q be quadratic over Ω_0 and let q_X^{asy} denote the asymptotic finite element approximation. Then, for all $\alpha \in \mathcal{A}_0$,*

$$\left.\begin{aligned} B_{C(\boldsymbol{c}_\alpha,h)}(q - q_X^{\mathrm{asy}}, v) &= 0 \quad \text{for all } v \in X_\alpha^{\mathrm{per}} \\ \int_{C(\boldsymbol{c}_\alpha,h)}(q - q_X^{\mathrm{asy}})\,\mathrm{d}\boldsymbol{x} &= 0. \end{aligned}\right\} \tag{7.35}$$

Furthermore, $q_X^{\mathrm{asy}} \in X(\Omega_0)$ and

$$\left.\begin{aligned} B_{\Omega_0}(q - q_X^{\mathrm{asy}}, v) &= 0 \quad \text{for all } v \in X_0(\Omega_0) \\ \int_{\Omega_0}(q - q_X^{\mathrm{asy}})\,\mathrm{d}\boldsymbol{x} &= 0. \end{aligned}\right\} \tag{7.36}$$

Proof. Observe that

$$q - q_X^{\mathrm{asy}} = q - I_X q - \widetilde{\Pi}^{\mathrm{per}}(q - I_X q)$$

and the first claim follows at once from (7.30). Now, by definition, there exists $\widehat{q} \in \widehat{X}^{\mathrm{per}}$ such that

$$(q - q_X^{\mathrm{asy}})|_{C(\boldsymbol{c}_\alpha,h)} = \widehat{q}(h^{-1}(\boldsymbol{x} - \boldsymbol{c}_\alpha)) \quad \boldsymbol{x} \in C(\boldsymbol{c}_\alpha, h).$$

Let $v \in X_0(\Omega_0)$ then direct use of the definitions gives

$$B_{\Omega_0}(q - q_X^{\mathrm{asy}}, v) = \sum_{\alpha \in \mathcal{A}_0} B_{C(\boldsymbol{c}_\alpha,h)}(q - q_X^{\mathrm{asy}}, v) = \widehat{B}(\widehat{q}, \widehat{v}) = 0$$

where

$$\widehat{v}(\widehat{\boldsymbol{x}}) = \sum_{\alpha \in \mathcal{A}_0} v(\boldsymbol{c}_\alpha + h\widehat{\boldsymbol{x}}) \in \widehat{X}^{\mathrm{per}}.$$

The second result is obtained by observing that

$$\begin{aligned} \int_{C(\boldsymbol{c}_\alpha,h)} q_X^{\mathrm{asy}}\,\mathrm{d}\boldsymbol{x} &= \int_{C(\boldsymbol{c}_\alpha,h)} I_X q\,\mathrm{d}\boldsymbol{x} + \int_{C(\boldsymbol{c}_\alpha,h)} \widetilde{\Pi}^{\mathrm{per}}(q - I_X q)\,\mathrm{d}\boldsymbol{x} \\ &= \int_{C(\boldsymbol{c}_\alpha,h)} I_X q\,\mathrm{d}\boldsymbol{x} + \int_{C(\boldsymbol{c}_\alpha,h)} (q - I_X q)\,\mathrm{d}\boldsymbol{x} \\ &= \int_{C(\boldsymbol{c}_\alpha,h)} q\,\mathrm{d}\boldsymbol{x} \end{aligned}$$

and then summing over the individual cells in Ω_0. ■

The next result is concerned with establishing local rates of convergence of the asymptotic finite element approximation over individual cells.

Lemma 7.9 *Suppose that condition (A1) holds and let q be as defined in (7.3). Then, for each cell $\alpha \in \mathcal{A}_0$, there holds*

$$ch^3 \leq \|q - q_X^{\text{asy}}\|_{L_2(\mathcal{C}(\boldsymbol{c}_\alpha,h))} \leq Ch^3 \tag{7.37}$$

and

$$ch^2 \leq |\!|\!| q - q_X^{\text{asy}} |\!|\!|_{\mathcal{C}(\boldsymbol{c}_\alpha,h)} \leq Ch^2 \tag{7.38}$$

where c and C are constants depending on μ_0 and $\|u\|_{W^{3,\infty}(\Omega_1)}$ but independent of h and α.

Proof. Firstly, observe that

$$q - q_X^{\text{asy}} = (q - I_X q) - \widetilde{\Pi}^{\text{per}}(q - I_X q)$$

and hence $q - q_X^{\text{asy}} \in \widetilde{H}^{1,\text{per}}(\Omega_0)$. As a consequence, the values of the norms of $q - q_X^{\text{asy}}$ are the same in every cell.

Let $\alpha \in \mathcal{A}_0$. Then, since both $q - q_X^{\text{asy}}$ and $q - I_X q$ belong to the space $\widetilde{X}^{\text{per}}$, it follows that the difference $I_X q - q_X^{\text{asy}}$ belongs to this space. Therefore, applying (7.35),

$$\begin{aligned} |\!|\!| q - q_X^{\text{asy}} |\!|\!|^2_{\mathcal{C}(\boldsymbol{c}_\alpha,h)} &= B_{\mathcal{C}(\boldsymbol{c}_\alpha,h)}(q - q_X^{\text{asy}}, q - I_X q) \\ &\leq |\!|\!| q - q_X^{\text{asy}} |\!|\!|_{\mathcal{C}(\boldsymbol{c}_\alpha,h)} |\!|\!| q - I_X q |\!|\!|_{\mathcal{C}(\boldsymbol{c}_\alpha,h)} \end{aligned}$$

and hence

$$|\!|\!| q - q_X^{\text{asy}} |\!|\!|_{\mathcal{C}(\boldsymbol{c}_\alpha,h)} \leq |\!|\!| q - I_X q |\!|\!|_{\mathcal{C}(\boldsymbol{c}_\alpha,h)} . \tag{7.39}$$

Applying the Aubin–Nitsche argument [50] over the convex domain $\mathcal{C}(\boldsymbol{c}_\alpha, h)$, one obtains

$$\|q - q_X^{\text{asy}}\|_{L_2(\mathcal{C}(\boldsymbol{c}_\alpha,h))} \leq Ch \, |\!|\!| q - q_X^{\text{asy}} |\!|\!|_{\mathcal{C}(\boldsymbol{c}_\alpha,h)} \leq Ch \, |\!|\!| q - I_X q |\!|\!|_{\mathcal{C}(\boldsymbol{c}_\alpha,h)} . \tag{7.40}$$

Therefore,

$$|\!|\!| q - q_X^{\text{asy}} |\!|\!|_{\mathcal{C}(\boldsymbol{c}_\alpha,h)} + h^{-1} \|q - q_X^{\text{asy}} q\|_{L_2(\mathcal{C}(\boldsymbol{c}_\alpha,h))} \leq C \, |\!|\!| q - I_X q |\!|\!|_{\mathcal{C}(\boldsymbol{c}_\alpha,h)} \leq Ch^2$$

thanks to standard approximation properties of the interpolant $I_X q$.

The lower bounds follow at once from Lemma 7.2 owing to the fact that the second-order derivatives of q are constant over the subdomain Ω_0. ■

Finally, we come to a key result relating the asymptotic finite element approximation with the Galerkin projection over the subdomain Ω_1.

Theorem 7.10 *Suppose conditions (A1) and (A2) hold and let* $\sigma = \frac{2}{5}(\tau - \frac{3}{2})$. *If the diameters* D_1 *and* D_2 *of the subdomains* $\Omega_2 \subset \Omega_1 \subset \Omega_0 \subset \Omega$ *satisfy*

$$ch^{(1+\sigma)/2} \leq D_2 \leq D_1 \leq Ch^{(1+\sigma)/2} \tag{7.41}$$

and the distance between the subdomains satisfies

$$d = \text{dist}(\partial\Omega_1, \partial\Omega_2) \geq Ch^{\sigma} D_1 \tag{7.42}$$

for positive constants c *and* C *independent of* h, *then*

$$|||q_X^{\text{asy}} - q_X^*|||_{\Omega_2} \leq Ch^{\sigma} \, |||e|||_{\Omega_2} \tag{7.43}$$

and

$$\left| |||e|||_{\Omega_2} - |||q - q_X^{\text{asy}}|||_{\Omega_2} \right| \leq Ch^{\sigma} \, |||e|||_{\Omega_2} \tag{7.44}$$

where C *is a positive constant depending on* $||u||_{W^{3,\infty}(\Omega_1)}$ *and* μ_0.

Proof. Let $v \in X_0(\Omega_0)$ with $\text{supp}(v) \subset \Omega_1$. Then, using the definition (7.4) of q_X^* and Lemma 7.8, we obtain

$$\begin{aligned} B_{\Omega_1}(q_X^* - q_X^{\text{asy}}, v) &= B_{\Omega_1}(q_X^*, v) - B_{\Omega_1}(q_X^{\text{asy}}, v) \\ &= B_{\Omega_1}(q, v) - B_{\Omega_1}(q, v) = 0. \end{aligned}$$

Moreover, $q_X^* - q_X^{\text{asy}} \in X(\Omega_1)$ and, as shown in the proof of Theorem 7.5, $d \geq Ch^{(1+3\sigma)/2}$. Therefore, applying Lemma 7.3 gives, on selecting $v = q_X^* - q_X^{\text{asy}}$,

$$|||q_X^* - q_X^{\text{asy}}|||_{\Omega_2} \leq \frac{C}{d} \, ||q_X^* - q_X^{\text{asy}}||_{L_2(\Omega_1)} \, .$$

Suppose for a moment that

$$||q_X^* - q_X^{\text{asy}}||_{L_2(\Omega_1)} \leq Ch \, |||q - I_X q|||_{\Omega_1} \, . \tag{7.45}$$

Then, applying approximation properties of the interpolant and the lower bound (7.20) leads to the conclusion

$$\begin{aligned} |||q - I_X q|||_{\Omega_1} &\leq Ch \, ||q||_{H^2(\Omega_1)} \\ &\leq ChD_1 \, ||q||_{W^{2,\infty}(\Omega_1)} \\ &\leq Ch^{1+(1+\sigma)/2-(3+\sigma)/2} \, |||e|||_{\Omega_2} \\ &= C \, |||e|||_{\Omega_2} \end{aligned}$$

where the constant depends on the function q. Consequently, thanks to the above estimates and the fact that $d \geq Ch^{(1+3\sigma)/2}$, we obtain

$$\begin{aligned} |||q_X^* - q_X^{\text{asy}}|||_{\Omega_2} &\leq \frac{Ch}{d} \, |||q - I_X q|||_{\Omega_1} \\ &\leq \frac{Ch}{d} \, |||e|||_{\Omega_2} \\ &\leq Ch^{(1-3\sigma)/2} \, |||e|||_{\Omega_2} \\ &\leq Ch^{\sigma} \, |||e|||_{\Omega_2} \end{aligned}$$

since $\sigma \leq 1/5$ implies that $(1-3\sigma)/2 \geq 1/5 \geq \sigma$.

It remains to prove that the estimate (7.45) holds. By the optimality property of the Galerkin projection, we have

$$|||q - q_X^*|||_{\Omega_1} \leq |||q - I_X q|||_{\Omega_1}$$

and from (7.39) we obtain

$$|||q - q_X^{\text{asy}}|||_{\Omega_1} \leq |||q - I_X q|||_{\Omega_1}.$$

Hence,

$$|||q_X^* - q_X^{\text{asy}}|||_{\Omega_1} \leq 2\,|||q - I_X q|||_{\Omega_1}.$$

Applying the Aubin–Nitsche duality argument [50] over the convex subdomain Ω_1, one obtains

$$||q - q_X^*||_{L_2(\Omega_1)} \leq Ch\,|||q - q_X^*|||_{\Omega_1} \leq Ch\,|||q - I_X q|||_{\Omega_1}$$

while by (7.40) we have

$$||q - q_X^{\text{asy}}||_{L_2(\Omega_1)} \leq Ch\,|||q - I_X q|||_{\Omega_1}$$

and so

$$||q_X^* - q_X^{\text{asy}}||_{L_2(\Omega_1)} \leq Ch\,|||q - I_X q|||_{\Omega_1}$$

which is precisely (7.45). The estimate (7.44) follows from Theorem 7.5. ■

7.5 STABILITY OF ESTIMATORS

Let K be any element in the partition, and let $\widetilde{K}$ denote the patch consisting of the element K and its neighboring elements. Suppose that we have a finite element approximation $u_X \in X$ of the solution to a problem with data f, and let the *a posteriori* error estimator corresponding to element K be denoted by $\epsilon(u_X, f, K)$. The following assumption will be made concerning the estimator:

(E) The estimator for the error over element K may be written in the form

$$\epsilon(u_X, f, K)^2 = \sum_{j,k=1}^{m} L_j(u_X, f, K) E_{jk} L_k(u_X, f, K) \tag{7.46}$$

where m is a positive integer, and the matrix $\boldsymbol{E}$, formed from the coefficients E_{jk}, is symmetric and nonnegative definite with a uniformly bounded largest eigenvalue,

$$\boldsymbol{E} = \boldsymbol{E}^\top; \quad \boldsymbol{a}^\top \boldsymbol{E} \boldsymbol{a} \leq C \boldsymbol{a}^\top \boldsymbol{a}, \quad \forall \boldsymbol{a} \in \mathbb{R}^m \tag{7.47}$$

and the functionals L_k, $k = 1, ..., m$, are subadditive,

$$L_k(u_X + v_X, f + g, K) \leq L_k(u_X, f, K) + L_k(v_X, g, K), \quad k \in \{1, \ldots, m\} \tag{7.48}$$

and stable in the sense that, for all $u_X \in X$ and $f \in L_\infty(K)$,

$$L_k(u_X, f, K) \leq C \left\{ |||u_X|||_{\widetilde{K}} + h^2 \|f\|_{L_\infty(\widetilde{K})} \right\} \tag{7.49}$$

where C is a positive constant that is independent of h. Finally, the estimator for the error over a subdomain Ω' consisting of elements $K \in \mathcal{P}'$ is obtained by summing contributions from each of the elements

$$\epsilon(u_X, f, \Omega')^2 = \sum_{K \in \mathcal{P}'} \epsilon(u_X, f, K)^2. \tag{7.50}$$

These assumptions are quite general and are satisfied by practically every estimator that is in common use.

The following sections illustrate how the condition may be verified for some particular estimators.

7.5.1 Verification of Stability Condition for Explicit Estimator

The explicit estimator for the error on element K is given by (see Chapter 2)

$$\epsilon(u_X, f, K)^2 = h_K^2 \|f + \Delta u_X\|_{L_2(K)}^2 + \frac{1}{2} \sum_{\gamma \subset \partial K} h_\gamma \left\| \left[\frac{\partial u_X}{\partial n} \right] \right\|_{L_2(\gamma)}^2 \tag{7.51}$$

where $[\cdot]$ denotes the jump in the normal fluxes on the interelement boundary. Define L_1 by

$$L_1(u_X, f, K) = h_K \|f + \Delta u_X\|_{L_2(K)} \tag{7.52}$$

and for each of the three edges γ_k, $k \in \{2, 3, 4\}$ of K, define L_k to be

$$L_k(u_X, f, K) = \left(\frac{h_{\gamma_k}}{2} \right)^{1/2} \left\| \left[\frac{\partial u_X}{\partial n} \right] \right\|_{L_2(\gamma_k)}. \tag{7.53}$$

Then,

$$\epsilon(u_X, f, K)^2 = \sum_{j,k=1}^{4} L_j(u_X, f, K) E_{jk} L_k(u_X, f, K) \tag{7.54}$$

where, in this particular case, $\boldsymbol{E}$ is the identity matrix. It is easily verified that each of the quantities L_k is subadditive. For instance, in the cases $k \in \{2, 3, 4\}$, the linearity of the jump operator $[\cdot]$ means that

$$L_k(u_X + v_X, f + g, K) = \left(\frac{h_{\gamma_k}}{2} \right)^{1/2} \left\| \left[\frac{\partial (u_X + v_X)}{\partial n} \right] \right\|_{L_2(\gamma_k)}$$

$$
\begin{aligned}
&= \left(\frac{h_{\gamma_k}}{2}\right)^{1/2} \left\| \left[\frac{\partial u_X}{\partial n}\right] + \left[\frac{\partial v_X}{\partial n}\right] \right\|_{L_2(\gamma_k)} \\
&\leq \left(\frac{h_{\gamma_k}}{2}\right)^{1/2} \left\| \left[\frac{\partial u_X}{\partial n}\right] \right\|_{L_2(\gamma_k)} \\
&+ \left(\frac{h_{\gamma_k}}{2}\right)^{1/2} \left\| \left[\frac{\partial v_X}{\partial n}\right] \right\|_{L_2(\gamma_k)} \\
&= L_k(u_X, f, K) + L_k(v_X, g, K).
\end{aligned}
$$

The proof is similar in the case of L_1. The stability of the functional L_1 is obtained by arguing as follows:

$$
\begin{aligned}
L_1(u_X, f, K) &= h_K \|f + \Delta u_X\|_{L_2(K)} \\
&\leq h_K \|f\|_{L_2(K)} + h_K \|\Delta u_X\|_{L_2(K)} \\
&\leq C \left\{ h_K^2 \|f\|_{L_\infty(K)} + \|u_X\|_K \right\}
\end{aligned}
$$

where an inverse estimate (see Theorem 1.3) has been applied to bound the second derivatives of u_X.

The stability of the functionals involving the jump in derivatives requires a preparatory result. Consider the jump on the edge γ shared by elements K and K':

$$
\left[\frac{\partial u_X}{\partial n}\right]\Bigg|_\gamma = \boldsymbol{n}_K \cdot \boldsymbol{\nabla} u_X|_K + \boldsymbol{n}_{K'} \cdot \boldsymbol{\nabla} u_X|_{K'}.
$$

Then, since the gradient of u_X is piecewise constant, we obtain

$$
\boldsymbol{n}_K \cdot \boldsymbol{\nabla} u_X|_K = \frac{1}{|K|} \boldsymbol{n}_K \cdot \int_K \boldsymbol{\nabla} u_X \, \mathrm{d}\boldsymbol{x}
$$

and so, by the Cauchy–Schwarz inequality, we have

$$
\begin{aligned}
\left| \boldsymbol{n}_K \cdot \boldsymbol{\nabla} u_X|_K \right| &\leq \frac{1}{|K|} \left| \int_K \boldsymbol{\nabla} u_X \, \mathrm{d}\boldsymbol{x} \right| \\
&\leq C h_K^{-1} \|\boldsymbol{\nabla} u_X\|_{L_2(K)} \\
&= C h_K^{-1} \|u_X\|_K,
\end{aligned}
$$

and hence it follows that for each edge γ, we obtain

$$
\left| \left[\frac{\partial u_X}{\partial n}\right]\Bigg|_\gamma \right| \leq C h_K^{-1} \|u_X\|_{\tilde{K}}. \tag{7.55}
$$

With the aid of this estimate and the fact that the meshes are locally quasi-uniform, it is easily seen that for all $k \in \{2, 3, 4\}$,

$$
L_k(u_X, f, K) \leq C h_\gamma \left| \left[\frac{\partial u_X}{\partial n}\right]\Bigg|_\gamma \right| \leq C \|u_X\|_{\tilde{K}}.
$$

Therefore, we conclude that the explicit estimator satisfies condition (E).

7.5.2 Verification of Stability Condition for Implicit Estimators

The implicit estimator is given by

$$\epsilon(u_X, f, K) = |||\phi_K|||_K \tag{7.56}$$

where $\phi_K \in V_K = \text{span}\{\theta_k : k = 1, \dots, m\}$ is the solution of the local residual boundary value problem,

$$B_K(\phi_K, v) = (f, v)_K - B_K(u_X, v) + \int_{\partial K} \left\langle \frac{\partial u_X}{\partial n_K} \right\rangle v \, \mathrm{d}s \quad \forall v \in V_K. \tag{7.57}$$

The space V_K consists of bubble functions as described in Section 3.3.2. Without loss of generality, the solution ϕ_K may be written in the form

$$\phi_K = \sum_{k=1}^{m} L_k(u_X, f, K)\theta_k \tag{7.58}$$

where the basis functions are normalised so that $|||\theta_k|||_K = 1$. The coefficients L_k are determined from the condition

$$\boldsymbol{E} \begin{bmatrix} L_1 \\ L_2 \\ \vdots \\ L_m \end{bmatrix} = \begin{bmatrix} r_1 \\ r_2 \\ \vdots \\ r_m \end{bmatrix} \tag{7.59}$$

where

$$r_k = (f, \theta_k)_K - B_K(u_X, \theta_k) + \int_{\partial K} \left\langle \frac{\partial u_X}{\partial n_K} \right\rangle \theta_k \, \mathrm{d}s \tag{7.60}$$

and $\boldsymbol{E}$ is the symmetric, nonnegative definite matrix whose entries are given by

$$E_{jk} = B_K(\theta_j, \theta_k) \leq |||\theta_j|||_K \, |||\theta_k|||_K = 1. \tag{7.61}$$

Thus, the largest eigenvalue of the matrix $\boldsymbol{E}$ is bounded by the dimension m of the space V_K. Furthermore, since the matrix $\boldsymbol{E}$ is invariant under a rescaling of the element K by $1/h_K$, it follows that the eigenvalues are independent of h_K. Consequently, the smallest eigenvalue of $\boldsymbol{E}$ is bounded away from zero by a constant that is independent of h_K. (The smallest eigenvalue is nonzero since, by assumption, the matrix is nonsingular.) It follows that the largest eigenvalue of $\boldsymbol{E}^{-1}$ is bounded above by a constant independent of h_K and thus

$$\max_{k \in \{1,\dots,m\}} L_k(u_X, f, K) \leq C \max_{k \in \{1,\dots,m\}} |r_k|.$$

Moreover, for any k,

$$\begin{aligned} |r_k| \;&\leq\; \|f\|_{L_2(K)} \|\theta_k\|_{L_2(K)} + |||u_X|||_K \, |||\theta_k|||_K \\ &\quad + \left\| \left\langle \frac{\partial u_X}{\partial n_K} \right\rangle \right\|_{L_2(\partial K)} \|\theta_k\|_{L_2(\partial K)} . \end{aligned}$$

Arguments similar to those used to arrive at (7.55) lead to the conclusion

$$\left| \left\langle \frac{\partial u_X}{\partial n} \right\rangle \Big|_{\gamma} \right| \leq C h_K^{-1} \, \| u_X \|_{\tilde{K}} ;$$

hence, since the length of the boundary ∂K is of order h_K, it follows that

$$\left\| \left\langle \frac{\partial u_X}{\partial n_K} \right\rangle \right\|_{L_2(\partial K)} \leq C h_K^{-1/2} \, \| u_X \|_{\tilde{K}} .$$

Furthermore, it is easily verified that

$$||\theta_k||_{L_2(K)} \leq C h_K$$

and

$$||\theta_k||_{L_2(\partial K)} \leq C h_K^{1/2}.$$

Consequently, gathering estimates gives

$$|r_k| \leq C \left\{ h_K^2 \, ||f||_{L_\infty(K)} + \| u_X \|_{\tilde{K}} \right\}$$

and the functionals L_k satisfy the local stability property. Finally, (7.59) shows that L_k are actually linear functionals, and thus subadditive. Therefore, the implicit estimator satisfies the condition (E).

7.5.3 Verification of Stability Condition for Recovery-Based Estimator

The recovery-based estimator is given by

$$\epsilon(u_X, f, K) = ||\nabla u_X - \boldsymbol{G}_X(u_X)||_{L_2(K)} \tag{7.62}$$

where $\boldsymbol{G}_X(u_X)$ is the recovered gradient. If the recovery operator $\boldsymbol{G}_X$ satisfies the boundedness and linearity condition $(R3)$ of Chapter 4, then it is straightforward to show the estimator satisfies condition (E), by choosing

$$L_1(u_X, f, K) = ||\nabla u_X - \boldsymbol{G}_X(u_X)||_{L_2(K)} . \tag{7.63}$$

The linearity property of $\boldsymbol{G}_X$ means that

$$\begin{aligned} & L_1(u_X + v_X, f + g, K) \\ = \; & ||\nabla(u_X + v_X) - \boldsymbol{G}_X(u_X + v_X)||_{L_2(K)} \\ = \; & ||\nabla u_X - \boldsymbol{G}_X(u_X) + \nabla v_X - \boldsymbol{G}_X(v_X)||_{L_2(K)} \\ \leq \; & ||\nabla u_X - \boldsymbol{G}_X(u_X)||_{L_2(K)} + ||\nabla v_X - \boldsymbol{G}_X(v_X)||_{L_2(K)} \\ = \; & L_1(u_X, f, K) + L_1(v_X, g, K) \end{aligned}$$

and thus L_1 is subadditive. Furthermore, the boundedness condition implies that

$$\begin{aligned} L_1(u_X, f, K) &\leq \|\boldsymbol{\nabla} u_X\|_{L_2(K)} + \|\boldsymbol{G}_X(u_X)\|_{L_2(K)} \\ &\leq |\!|\!| u_X |\!|\!|_K + Ch_K \|\boldsymbol{G}_X(u_X)\|_{L_\infty(K)} \\ &\leq |\!|\!| u_X |\!|\!|_K + Ch_K |u_X|_{W^{1,\infty}(\widetilde{K})}, \end{aligned}$$

then using equivalence of norms and a scaling argument reveals

$$|u_X|_{W^{1,\infty}(\widetilde{K})} \leq Ch_{\widetilde{K}}^{-1} |\!|\!| u_X |\!|\!|_{\widetilde{K}}$$

and it follows that

$$L_1(u_X, f, K) \leq C |\!|\!| u_X |\!|\!|_{\widetilde{K}}.$$

Therefore, the recovery-based estimator satisfies condition (E).

7.5.4 Elementary Consequences of the Stability Condition

Some elementary inferences of condition (E) will now be drawn. The first result shows that the estimator satisfies triangle inequalities.

Lemma 7.11 *Let ϵ satisfy condition (E). Then, for all u, $v \in X$ and f, $g \in L_\infty(\widetilde{K})$, there holds*

$$\epsilon(u+v, f+g, K) \leq \epsilon(u, f, K) + \epsilon(v, g, K) \tag{7.64}$$

and

$$|\epsilon(u, f, K) - \epsilon(v, g, K)| \leq \epsilon(u-v, f-g, K) \tag{7.65}$$

Proof. The second estimate is an immediate consequence of the first in the same way as the reverse triangle inequality follows from the standard triangle inequality. Thus, it is sufficient to prove the first estimate. Condition (E) shows that

$$\epsilon(u+v, f+g, K)^2 = \sum_{j,k=1}^{m} L_j(u+v, f+g, K) E_{jk} L_k(u+v, f+g, K)$$

and then the subadditivity assumption

$$L_k(u+v, f+g, K) \leq L_k(u, f, K) + L_k(v, g, K)$$

leads to

$$\begin{aligned} \epsilon(u+v, f+g, K)^2 &\leq \epsilon(u, f, K)^2 + \epsilon(v, g, K)^2 \\ &\quad + 2\sum_{j,k=1}^{m} L_j(u, f, K) E_{jk} L_k(v, g, K). \end{aligned}$$

The fact that the matrix $\boldsymbol{E}$ is symmetric and nonnegative implies, with aid of the Cauchy–Schwarz inequality, that

$$\sum_{j,k=1}^{m} L_j(u,f,K)E_{jk}L_k(v,g,K) \leq \epsilon(u,f,K)\epsilon(v,g,K)$$

and so

$$\begin{aligned}\epsilon(u+v,f+g,K)^2 &\leq \epsilon(u,f,K)^2 + 2\epsilon(u,f,K)\epsilon(v,g,K) + \epsilon(v,g,K)^2 \\ &= (\epsilon(u,f,K) + \epsilon(v,g,K))^2\end{aligned}$$

and the first estimate follows at once thanks to the nonnegativity of the estimator. ■

The next result shows that the estimator over an element K depends continuously on the data sampled over the patch $\widetilde{K}$.

Lemma 7.12 *Suppose that the estimator ϵ satisfies condition (E). Then, there exists a positive constant C such that for all u, $v \in X$ and f, $g \in L_\infty(\widetilde{K})$,*

$$|\epsilon(u,f,K) - \epsilon(v,g,K)| \leq C\left\{|||u-v|||_{\widetilde{K}} + h^2 \|f-g\|_{L_\infty(\widetilde{K})}\right\} \tag{7.66}$$

where C is a positive constant that depends on m, but is independent of h and K. Furthermore, let Ω' denote the subdomain composed from elements $K \in \mathcal{P}'$. Then

$$|\epsilon(u,f,\Omega') - \epsilon(v,g,\Omega')| \leq C\sum_{K\in\mathcal{P}'}\left\{|||u-v|||_{\widetilde{K}} + h^2 \|f-g\|_{L_\infty(\widetilde{K})}\right\} \tag{7.67}$$

where C is a positive constant that depends on m but is independent of h.

Proof. With the aid of Lemma 7.11,

$$|\epsilon(u,f,K) - \epsilon(v,g,K)| \leq \epsilon(u-v,f-g,K).$$

Applying parts (7.47) and (7.49) of the condition (E), then gives

$$\begin{aligned}\epsilon(u-v,f-g,K)^2 &\leq C\sum_{k=1}^{m} L_k(u-v,f-g,K)^2 \\ &\leq Cm\left\{|||u-v|||_{\widetilde{K}} + h^2\|f-g\|_{L_\infty(\widetilde{K})}\right\}^2\end{aligned}$$

and the proof of the first estimate is complete thanks to the nonnegativity of ϵ. The elementary inequality

$$\begin{aligned}&\left|\left\{\sum_{K\in\mathcal{P}'}\epsilon(u,f,K)^2\right\}^{1/2} - \left\{\sum_{K\in\mathcal{P}'}\epsilon(v,g,K)^2\right\}^{1/2}\right| \\ &\leq \left\{\sum_{K\in\mathcal{P}'}|\epsilon(u,f,K)-\epsilon(v,g,K)|^2\right\}^{1/2}\end{aligned}$$

means that the second estimate is an immediate consequence of the first. ■

7.5.5 Evaluation of Effectivity Index in the Asymptotic Limit

The next result shows that the difference between the estimates obtained by applying the estimator ϵ to the true data (u_X, f, Ω_2) and the model data $(q_X^{\text{asy}}, -\Delta q, \Omega_2)$ is, asymptotically in h, negligible in comparison with the true error over the subdomain.

Theorem 7.13 *Suppose conditions (A1) and (A2) hold, let $\sigma = \frac{2}{5}(\tau - \frac{3}{2})$ and let ϵ be an estimator satisfying condition (E). If the diameters D_1 and D_2 of the subdomains $\Omega_2 \subset \Omega_1 \subset \Omega_0 \subset \Omega$ satisfy*

$$ch^{(1+\sigma)/2} \leq D_2 \leq D_1 \leq Ch^{(1+\sigma)/2} \tag{7.68}$$

and the distance between the subdomains satisfies

$$d = \text{dist}(\partial\Omega_1, \partial\Omega_2) \geq Ch^\sigma D_1 \tag{7.69}$$

for positive constants c and C independent of h, then, for h sufficiently small,

$$|\epsilon(u_X, f, \Omega_2) - \epsilon(q_X^{\text{asy}}, -\Delta q, \Omega_2)| \leq Ch^\sigma \, \|\!|e|\!\|_{\Omega_2} \tag{7.70}$$

where C is a positive constant C depending on $\|u\|_{W^{3,\infty}(\Omega_1)}$ and μ_0.

Proof. Let Ω_2 be as in the statement of the Theorem and let $\widetilde{\Omega}_2$ denote the subdomain consisting of Ω_2 itself along with the single layer of cells that border Ω_2. The diameter $\widetilde{H}_2$ of the enlarged domain $\widetilde{\Omega}_2$ satisfies $D_2 \leq \widetilde{H}_2 \leq D_2 + Ch$; and as a consequence it follows that, for appropriate positive constants c and C independent of h,

$$ch^{(1+\sigma)/2} \leq \widetilde{H}_2 \leq Ch^{(1+\sigma)/2}.$$

Furthermore, since $d \geq Ch^\sigma D_1 = Ch^{(1+3\sigma)/2}$, it follows that for sufficiently small h, it may be assumed that

$$\text{dist}(\partial\Omega_1, \partial\widetilde{\Omega}_2) \geq Ch^{(1+3\sigma)/2} \geq Ch^\sigma D_1.$$

This shows that we could apply Theorem 7.4 to the pair of subdomains Ω_1 and $\widetilde{\Omega}_2$ and hence conclude that

$$\|\!|e|\!\|_{\widetilde{\Omega}_2} \leq Ch^{(3+\sigma)/2}$$

and then, thanks to Theorem 7.4,

$$Ch^{(3+\sigma)/2} \leq \|\!|e|\!\|_{\Omega_2}$$

and we obtain

$$\|\!|e|\!\|_{\widetilde{\Omega}_2} \leq C \, \|\!|e|\!\|_{\Omega_2}. \tag{7.71}$$

Applying Lemma 7.12 reveals that

$$\begin{aligned} &|\epsilon(u_X, f, \Omega_2) - \epsilon(q_X^{\text{asy}}, -\Delta q, \Omega_2)| \\ &\quad \leq C\left\{ |\!|\!| u_X - q_X^{\text{asy}} |\!|\!|_{\tilde{\Omega}_2} + h^2 \|f + \Delta q\|_{L_\infty(\tilde{\Omega}_2)} \right\}. \end{aligned}$$

Thanks to property (7.3) and (7.20), we have

$$\begin{aligned} h^2 \|f + \Delta q\|_{L_\infty(\tilde{\Omega}_2)} &\leq h^2 |u - q|_{W^{2,\infty}(\tilde{\Omega}_2)} \\ &\leq C D_1 h^2 \|u\|_{W^{3,\infty}(\tilde{\Omega}_2)} \\ &\leq C h^{2+(1+\sigma)/2-(3+\sigma)/2} |\!|\!| e |\!|\!|_{\tilde{\Omega}_2} \\ &\leq C h |\!|\!| e |\!|\!|_{\tilde{\Omega}_2} \end{aligned}$$

while by Theorem 7.5, Theorem 7.10, (7.3), and (7.20) we obtain

$$\begin{aligned} &|\!|\!| u_X - q_X^{\text{asy}} |\!|\!|_{\tilde{\Omega}_2} \\ &\quad \leq |\!|\!| (u - u_X) - (q - q_X^{\text{asy}}) |\!|\!|_{\tilde{\Omega}_2} + |\!|\!| u - q |\!|\!|_{\tilde{\Omega}_2} \\ &\quad \leq |\!|\!| (u - u_X) - (q - q_X^*) |\!|\!|_{\tilde{\Omega}_2} + |\!|\!| q_X^{\text{asy}} - q_X^* |\!|\!|_{\tilde{\Omega}_2} + |\!|\!| u - q |\!|\!|_{\tilde{\Omega}_2} \\ &\quad \leq C h^\sigma |\!|\!| e |\!|\!|_{\tilde{\Omega}_2} + C D_1^3 \|u\|_{W^{3,\infty}(\tilde{\Omega}_2)} \\ &\quad \leq C h^\sigma |\!|\!| e |\!|\!|_{\tilde{\Omega}_2} + C h^{3(1+\sigma)/2-(3+\sigma)/2} |\!|\!| e |\!|\!|_{\tilde{\Omega}_2} \\ &\quad = C h^\sigma |\!|\!| e |\!|\!|_{\tilde{\Omega}_2} \end{aligned}$$

and the result follows from (7.71). ■

The following result shows that, in the limit as $h \to 0$, the effectivity index of the estimator $\epsilon(u_X, f, \Omega_2)$ for the true error $|\!|\!| e |\!|\!|_{\Omega_2}$ is identical to the effectivity index of the estimator $\epsilon(q_X^{\text{asy}}, -\Delta q, \Omega_2)$ for the error in the asymptotic finite element approximation $|\!|\!| q - q_X^{\text{asy}} |\!|\!|_{\Omega_2}$.

Theorem 7.14 *Suppose conditions (A1) and (A2) hold, let $\sigma = \frac{2}{5}(\tau - \frac{3}{2})$, and suppose the diameters D_1 and D_2 of the subdomains $\Omega_2 \subset \Omega_1 \subset \Omega_0 \subset \Omega$ satisfy*

$$c h^{(1+\sigma)/2} \leq D_2 \leq D_1 \leq C h^{(1+\sigma)/2} \tag{7.72}$$

and the distance between the subdomains satisfies

$$d = \text{dist}(\partial\Omega_1, \partial\Omega_2) \geq C h^\sigma D_1 \tag{7.73}$$

for positive constants c and C independent of h. If ϵ is an error estimator satisfying condition (E) and if

$$\frac{\epsilon(q_X^{\text{asy}}, -\Delta q, \Omega_2)}{|\!|\!| q - q_X^{\text{asy}} |\!|\!|_{\Omega_2}} \leq \overline{C}, \tag{7.74}$$

then

$$\left| \frac{\epsilon(u_X, f, \Omega_2)}{|\!|\!| e |\!|\!|_{\Omega_2}} - \frac{\epsilon(q_X^{\text{asy}}, -\Delta q, \Omega_2)}{|\!|\!| q - q_X^{\text{asy}} |\!|\!|_{\Omega_2}} \right| \leq C h^\sigma \tag{7.75}$$

where C is a positive constant C depending on $\|u\|_{W^{3,\infty}(\Omega_1)}$, $\overline{C}$ and μ_0.

Proof. The quantity on the left-hand side of (7.75) may be bounded by

$$\frac{1}{\|e\|_{\Omega_2}}\left\{|\epsilon(u_X, f, \Omega_2) - \epsilon(q_X^{\mathrm{asy}}, -\Delta q, \Omega_2)| + \overline{C}\left|\|e\|_{\Omega_2} - \|q - q_X^{\mathrm{asy}}\|_{\Omega_2}\right|\right\};$$

therefore, applying Theorem 7.13 and Theorem 7.10 gives the result claimed. ■

Finally, we come to the key result of the whole chapter showing that in the limit as $h \to 0$, the effectivity index of the estimator $\epsilon(u_X, f, \Omega_2)$ for the true error $\|e\|_{\Omega_2}$ may be obtained by evaluating the effectivity index of the estimator for the error in the asymptotic finite element approximation over a single cell.

Corollary 7.15 *Let the assumptions of Theorem 7.14 hold. Then,*

$$\lim_{h\to 0}\frac{\epsilon(u_X, f, \Omega_2)}{\|e\|_{\Omega_2}} = \lim_{h\to 0}\frac{\epsilon(q_X^{\mathrm{asy}}, -\Delta q, C(\mathbf{c}_\alpha, h))}{\|q - q_X^{\mathrm{asy}}\|_{C(\mathbf{c}_\alpha, h)}} \tag{7.76}$$

where $C(\mathbf{c}_\alpha, h)$ is any cell contained in Ω_2.

Proof. The result follows at once from Theorem 7.14 since $q - q_X^{\mathrm{asy}}$ is periodic over Ω_2 and as a consequence the local effectivity index in each cell is the same. ■

7.6 AN APPLICATION OF THE THEORY

7.6.1 Computation of Asymptotic Finite Element Solution

The details of the procedure involved in computing the asymptotic finite element approximation over the mesh shown in Figure 7.1 will be illustrated for the case when the quadratic approximation q is a monomial $q \in \{x^2, xy, y^2\}$. The associated reference cell is the perturbed criss-cross pattern with the central node offset as shown in Figure 7.5. The computation is based on the identity (7.34),

$$q_X^{\mathrm{asy}} = I_X q + \tilde{z}^{\mathrm{per}} \tag{7.77}$$

where

$$\tilde{z}^{\mathrm{per}} = \tilde{\Pi}^{\mathrm{per}}(q - I_X q). \tag{7.78}$$

Lemma 7.6 then shows that

$$\tilde{z}^{\mathrm{per}}(\boldsymbol{x}) = \hat{z}\left(h^{-1}(\boldsymbol{x} - \mathbf{c}_\alpha)\right), \quad \boldsymbol{x} \in C(\mathbf{c}_\alpha, h) \tag{7.79}$$

where

$$\hat{z} = \hat{\Pi}^{\mathrm{per}}(\hat{q} - \hat{I}\hat{q}) \tag{7.80}$$

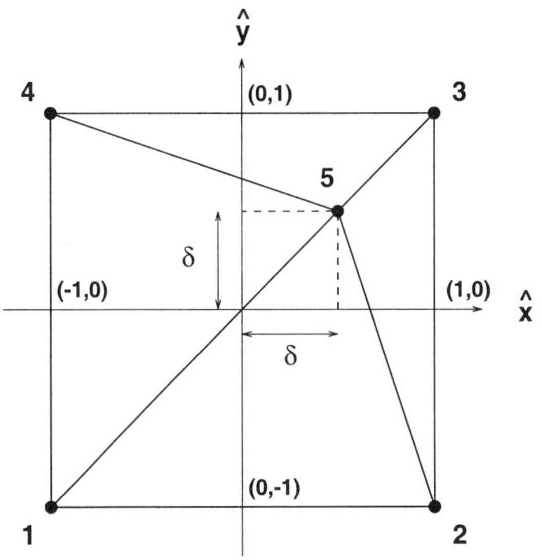

Fig. 7.5 Reference cell used in computation of asymptotic finite element solution over the mesh shown in Figure 7.1. The central node is located at (δ, δ).

with $\widehat{q}(\widehat{\boldsymbol{x}}) = q(\boldsymbol{c}_\alpha + h\widehat{\boldsymbol{x}})$ and where $\widehat{I\widehat{q}}$ is the piecewise linear interpolant at the vertices of the mesh on the reference cell. It follows that

$$\widehat{z} = h^2 \widehat{\Pi}^{\mathrm{per}}(w - \widehat{I}w) \tag{7.81}$$

where w is either $\widehat{x}^2$, $\widehat{x}\widehat{y}$, or $\widehat{y}^2$, depending on the original choice of q. Consequently, the problem of computing the asymptotic finite element approximation reduces to the construction of the periodic finite element approximation of $w - \widehat{I}w$ over the reference cell.

The periodic finite element subspace $\widehat{X}^{\mathrm{per}}$ associated with the reference cell shown in Figure 7.5 consists of two degrees of freedom: The periodicity condition means that the values of the approximation at the corners of the reference cell must agree. Thus, the space has one degree of freedom associated with the value of the approximation at the corners and a second degree of freedom associated with the value of the approximation at the central node.

A basis $\{\theta_1^{\mathrm{per}}, \theta_2^{\mathrm{per}}\}$ for the periodic subspace $\widehat{X}^{\mathrm{per}}$ may be constructed in terms of the standard Lagrange basis functions $\{\theta_1, \theta_2, \theta_3, \theta_4, \theta_5\}$ associated with the nodes shown in Figure 7.5. That is, each basis function θ_k is piecewise affine on the mesh with the values at the nodes given by $\theta_k(\boldsymbol{x}_\ell) = 1$ if $k = \ell$ and zero otherwise. A basis for the periodic space is then

$$\left.\begin{aligned} \theta_1^{\mathrm{per}} &= \theta_1 + \theta_2 + \theta_3 + \theta_4 \\ \theta_2^{\mathrm{per}} &= \theta_5, \end{aligned}\right\} \tag{7.82}$$

or, alternatively, in matrix form

$$\boldsymbol{\theta}^{\text{per}} = \boldsymbol{R}\boldsymbol{\theta} \tag{7.83}$$

where $\boldsymbol{R}$ is the matrix

$$\boldsymbol{R} = \begin{bmatrix} 1 & 1 & 1 & 1 & 0 \\ 0 & 0 & 0 & 0 & 1 \end{bmatrix}. \tag{7.84}$$

Let $\boldsymbol{B}$ denote the stiffness matrix that would be obtained by performing the standard finite element subassembly procedure on the unconstrained finite element subspace corresponding to the basis functions $\{\theta_k\}$. The stiffness matrix $\boldsymbol{B}^{\text{per}}$ associated with the periodic finite element approximation is found, with the aid of (7.83), to be

$$\boldsymbol{B}^{\text{per}} = \boldsymbol{R}\boldsymbol{B}\boldsymbol{R}^{\top}. \tag{7.85}$$

As a matter of fact, the change of basis could be performed more efficiently at the element level before the subassembly procedure. However, efficiency is a minor issue since periodic cell computations are generally of low dimension.

For the particular mesh shown in Figure 7.5, the unconstrained stiffness matrix is found to be

$$\boldsymbol{B} = \begin{bmatrix} \frac{1+\delta^2}{1+\delta} & -\frac{\delta}{2} & 0 & -\frac{\delta}{2} & -\frac{1-\delta}{1+\delta} \\ -\frac{\delta}{2} & 1 & \frac{\delta}{2} & 0 & -1 \\ 0 & \frac{\delta}{2} & \frac{1+\delta^2}{1-\delta} & \frac{\delta}{2} & -\frac{1+\delta}{1-\delta} \\ -\frac{\delta}{2} & 0 & \frac{\delta}{2} & 1 & -1 \\ -\frac{1-\delta}{1+\delta} & -1 & -\frac{1+\delta}{1-\delta} & -1 & \frac{4}{(1+\delta)(1-\delta)} \end{bmatrix} \tag{7.86}$$

and the stiffness matrix for the periodic cell problem is

$$\boldsymbol{B}^{\text{per}} = \frac{4}{(1+\delta)(1-\delta)} \begin{bmatrix} 1 & -1 \\ -1 & 1 \end{bmatrix}. \tag{7.87}$$

As is to be expected, this matrix is singular with the null space corresponding to the constant mode. The constant mode is determined by the additional condition imposed on the average value of the periodic finite element solution.

The load vector for the periodic problem is

$$\boldsymbol{f}^{\text{per}} = \widehat{B}(w - \widehat{I}w, \boldsymbol{\theta}^{\text{per}}). \tag{7.88}$$

As for the computation of the stiffness matrix, it is convenient to express this in terms of the load vector $\boldsymbol{f}$ corresponding to the unconstrained basis functions

$$\boldsymbol{f} = \widehat{B}(w - \widehat{I}w, \boldsymbol{\theta}) \tag{7.89}$$

using the change of basis (7.83),

$$\boldsymbol{f}^{\text{per}} = \boldsymbol{R}\boldsymbol{f}. \tag{7.90}$$

The computation of the load vector relative to the unconstrained basis is based on the formula

$$\begin{aligned} \boldsymbol{f} &= \widehat{B}(w, \boldsymbol{\theta}) - \widehat{B}(\widehat{I}w, \boldsymbol{\theta}) \\ &= \widehat{B}(w, \boldsymbol{\theta}) - \boldsymbol{B}\boldsymbol{w}^{\text{int}} \end{aligned}$$

where $\boldsymbol{w}^{\text{int}}$ is the vector formed using the values of w at the vertices of the reference cell,

$$\boldsymbol{w}^{\text{int}} = \begin{bmatrix} q(\widehat{\boldsymbol{x}}_1) \\ q(\widehat{\boldsymbol{x}}_2) \\ q(\widehat{\boldsymbol{x}}_3) \\ q(\widehat{\boldsymbol{x}}_4) \\ q(\widehat{\boldsymbol{x}}_5) \end{bmatrix}. \tag{7.91}$$

For the cell shown in Figure 7.5, one finds that

$$\boldsymbol{w}^{\text{int}} = \begin{bmatrix} 1 \\ 1 \\ 1 \\ 1 \\ \delta^2 \end{bmatrix}, \begin{bmatrix} 1 \\ -1 \\ 1 \\ -1 \\ \delta^2 \end{bmatrix}, \begin{bmatrix} 1 \\ 1 \\ 1 \\ 1 \\ \delta^2 \end{bmatrix} \tag{7.92}$$

for the cases $w = \widehat{x}^2$, $\widehat{x}\widehat{y}$, and $\widehat{y}^2$, respectively. Furthermore,

$$\boldsymbol{B}\boldsymbol{w}^{\text{int}} = \begin{bmatrix} (1-\delta)^2 \\ 1-\delta^2 \\ (1+\delta)^2 \\ 1-\delta^2 \\ -4 \end{bmatrix}, \begin{bmatrix} 1+\delta^2 \\ -(1+\delta^2) \\ 1+\delta^2 \\ -(1+\delta^2) \\ 0 \end{bmatrix}, \begin{bmatrix} (1-\delta)^2 \\ 1-\delta^2 \\ (1+\delta)^2 \\ 1-\delta^2 \\ -4 \end{bmatrix} \tag{7.93}$$

and

$$\widehat{B}(w, \boldsymbol{\theta}) = \frac{2}{3}\begin{bmatrix} 1-2\delta \\ 1 \\ 1+2\delta \\ 1 \\ -4 \end{bmatrix}, \frac{2}{3}\begin{bmatrix} 1 \\ -1 \\ 1 \\ -1 \\ 0 \end{bmatrix}, \frac{2}{3}\begin{bmatrix} 1-2\delta \\ 1 \\ 1+2\delta \\ 1 \\ -4 \end{bmatrix} \tag{7.94}$$

so that

$$\boldsymbol{f} = -\frac{1}{3}\begin{bmatrix} 1-2\delta+3\delta^2 \\ 1-3\delta^2 \\ 1+2\delta+3\delta^2 \\ 1-3\delta^2 \\ -4 \end{bmatrix}, \quad -\frac{1}{3}\begin{bmatrix} 1+3\delta^2 \\ -(1+3\delta^2) \\ 1+3\delta^2 \\ -(1+3\delta^2) \\ 0 \end{bmatrix}, \quad -\frac{1}{3}\begin{bmatrix} 1-2\delta+3\delta^2 \\ 1-3\delta^2 \\ 1+2\delta+3\delta^2 \\ 1-3\delta^2 \\ -4 \end{bmatrix} \tag{7.95}$$

and

$$\boldsymbol{f}^{\text{per}} = \frac{4}{3}\begin{bmatrix} -1 \\ 1 \end{bmatrix}, \quad \begin{bmatrix} 0 \\ 0 \end{bmatrix}, \quad \frac{4}{3}\begin{bmatrix} -1 \\ 1 \end{bmatrix}. \tag{7.96}$$

The general solution of the system

$$\boldsymbol{B}^{\text{per}}\boldsymbol{w} = \boldsymbol{f}^{\text{per}} \tag{7.97}$$

is given by

$$\boldsymbol{w} = -\frac{1}{9}\begin{bmatrix} 4 + 2\delta^2 + c_1 \\ 1 + 5\delta^2 + c_1 \end{bmatrix}, \quad \begin{bmatrix} 3\delta^2 + c_2 \\ 3\delta^2 + c_2 \end{bmatrix}, \quad -\frac{1}{9}\begin{bmatrix} 4 + 2\delta^2 + c_3 \\ 1 + 5\delta^2 + c_3 \end{bmatrix} \tag{7.98}$$

where c_1, c_2, and c_3 are arbitrary constants. The periodic finite element approximation over the reference cell is given by

$$\widehat{\Pi}^{\text{per}}(w - \widehat{I}w) = \boldsymbol{w}^\top \boldsymbol{\theta}^{\text{per}} = \boldsymbol{w}^\top \boldsymbol{R}\boldsymbol{\theta} = \left(\boldsymbol{R}^\top \boldsymbol{w}\right)^\top \boldsymbol{\theta}$$

and the undetermined constants are fixed by the requirement

$$\int_{\mathcal{C}(\mathbf{0},1)} (w - \widehat{I}w)\,\mathrm{d}\boldsymbol{x} = \int_{\mathcal{C}(\mathbf{0},1)} \widehat{\Pi}^{\text{per}}(w - \widehat{I}w)\,\mathrm{d}\boldsymbol{x} \tag{7.99}$$

that arises from the definition (7.29). In particular, for the cases considered above, a simple computation reveals that $c_1 = c_2 = c_3 = 0$.

The function $\widehat{z}$ introduced in (7.80) is piecewise linear on the mesh in the reference cell with the value at the vertices of the reference cell given by $h^2\boldsymbol{R}^\top\boldsymbol{w}$. In particular, for the cases $q = x^2$, xy, and y^2, the value of $\widehat{z}$ at the vertices are, respectively,

$$-\frac{h^2}{9}\begin{bmatrix} 4 + 2\delta^2 \\ 4 + 2\delta^2 \\ 4 + 2\delta^2 \\ 4 + 2\delta^2 \\ 1 + 5\delta^2 \end{bmatrix}, \quad h^2\begin{bmatrix} 3\delta^2 \\ 3\delta^2 \\ 3\delta^2 \\ 3\delta^2 \\ 3\delta^2 \end{bmatrix}, \quad -\frac{h^2}{9}\begin{bmatrix} 4 + 2\delta^2 \\ 4 + 2\delta^2 \\ 4 + 2\delta^2 \\ 4 + 2\delta^2 \\ 1 + 5\delta^2 \end{bmatrix}. \tag{7.100}$$

In turn, the asymptotic finite element approximation q_X^{asy} is piecewise linear on each mesh cell $\mathcal{C}(\boldsymbol{c}_\alpha, h)$ with values at the node located at $\boldsymbol{x}'$ given by

$$q(\boldsymbol{x}') + \widehat{z}\left(h^{-1}(\boldsymbol{x}' - \boldsymbol{c}_\alpha)\right). \tag{7.101}$$

7.6.2 Evaluation of the Error in Asymptotic Finite Element Approximation

The error in the approximation of the monomial $q_\beta = \boldsymbol{x}^\beta$ by the asymptotic finite element approximation is given by

$$q_\beta - (q_\beta)_X^{\text{asy}} = q_\beta - I_X q_\beta - \widetilde{z}_\beta^{\text{per}} \tag{7.102}$$

where, similarly to (7.78),

$$\tilde{z}_\beta^{\text{per}} = h^2 \widehat{\Pi}^{\text{per}}(w_\beta - \widehat{I} w_\beta). \tag{7.103}$$

with $w_\beta = \widehat{\boldsymbol{x}}^\beta$. Moreover, with the aid of Lemma 7.7, it follows that

$$q_\beta - I_X q_\beta = h^2(w_\beta - \widehat{I} w_\beta) \tag{7.104}$$

and hence,

$$q_\beta - (q_\beta)_X^{\text{asy}} = h^2 \left((w_\beta - \widehat{I} w_\beta) - \Pi^{\text{per}}(w_\beta - \widehat{I} w_\beta) \right). \tag{7.105}$$

This identity suggests defining the asymptotic finite element approximation w_β^{asy} over the reference cell, by analogy with (7.34), to be

$$w_\beta^{\text{asy}} = \widehat{I} w_\beta + \widehat{\Pi}^{\text{per}}(w_\beta - \widehat{I} w_\beta), \tag{7.106}$$

so that

$$q_\beta - (q_\beta)_X^{\text{asy}} = h^2(w_\beta - w_\beta^{\text{asy}}). \tag{7.107}$$

This shows how the error in the asymptotic finite element approximation of q_β over a physical cell is related to the error in the asymptotic finite element approximation of w_β over the reference cell.

If we wish to approximate a quadratic polynomial of the form

$$q(\boldsymbol{x}) = \sum_{|\beta|=2} a_\beta \boldsymbol{x}^\beta = \sum_{|\beta|=2} a_\beta w_\beta, \tag{7.108}$$

then linearity implies that the asymptotic finite element approximation of q is given by

$$q_X^{\text{asy}} = \sum_{|\beta|=2} a_\beta \,(q_\beta)_X^{\text{asy}} \tag{7.109}$$

or, with the aid of (7.107),

$$q_X^{\text{asy}} = h^2 \sum_{|\beta|=2} a_\beta w_\beta^{\text{asy}}. \tag{7.110}$$

It follows that

$$|\!|\!| q - q_X^{\text{asy}} |\!|\!|^2_{C(\boldsymbol{c}_\alpha, h)} = h^4 \sum_{|\beta|=2} \sum_{|\gamma|=2} a_\beta \mathcal{E}_{\beta\gamma} a_\gamma \tag{7.111}$$

where the entries $\mathcal{E}_{\beta\gamma}$ are given by

$$\mathcal{E}_{\beta\gamma} = \widehat{B}(w_\beta - w_\beta^{\text{asy}}, w_\gamma - w_\gamma^{\text{asy}}). \tag{7.112}$$

The earlier computations leading to the asymptotic finite element approximations of the monomials x^2, xy, and y^2 are then used to deduce that

$$\mathcal{E} = \frac{4}{9} \begin{bmatrix} 4(2+\delta^2) & 6\delta^2 & -4(1-\delta^2) \\ 6\delta^2 & 3(1+3\delta^2) & 6\delta^2 \\ -4(1-\delta^2) & 6\delta^2 & 4(2+\delta^2) \end{bmatrix}. \tag{7.113}$$

7.6.3 Computation of Limits on the Asymptotic Effectivity Index for Zienkiewicz–Zhu Patch Recovery Estimator

Let ϵ be an *a posteriori* error estimator whose performance we wish to investigate. Corollary 7.15 shows that the effectivity index of the estimator over an interior subdomain consisting of a translation invariant mesh can be obtained by performing computations over a single cell in the subdomain for each particular set of data u_X and f corresponding to a particular finite element approximation. In this section, it will be shown how limits on the asymptotic values of the effectivity index may be obtained for *all possible* sets of data satisfying the assumptions $(A1)$ and $(A2)$.

To begin with, consider once again the special case where $q_\beta = \boldsymbol{x}^\beta$ and let $(q_\beta)_X^{\text{asy}}$ be the asymptotic finite element approximation, and for the sake of definiteness, suppose that ϵ is taken to be the recovery-based estimator of the form

$$\epsilon(u_X, f, K) = \|\boldsymbol{G}(u_X) - \boldsymbol{\nabla} u_X\|_{L_2(K)}, \tag{7.114}$$

where $\boldsymbol{G}$ is a recovery operator as in Chapter 4, and K is an element. Thus, applying the estimator to the approximation of q_β by $(q_\beta)_X^{\text{asy}}$ over any of the physical cells $\mathcal{C}(\boldsymbol{c}_\alpha, h)$ yields the estimator

$$\epsilon\left((q_\beta)_X^{\text{asy}}, -\Delta q, \mathcal{C}(\boldsymbol{c}_\alpha, h)\right)^2 = \sum_{K \subset \mathcal{C}(\boldsymbol{c}_\alpha, h)} \|\boldsymbol{G}\left((q_\beta)_X^{\text{asy}}\right) - \boldsymbol{\nabla}(q_\beta)_X^{\text{asy}}\|_{L_2(K)}^2 . \tag{7.115}$$

Alternatively, the estimator could be applied to the asymptotic finite element approximation w_β^{asy} of the mapped monomial w_β over the reference cell, giving

$$\epsilon\left(w_\beta^{\text{asy}}, -\Delta w_\beta, \mathcal{C}(\boldsymbol{0}, 1)\right)^2 = \sum_{K \subset \mathcal{C}(\boldsymbol{0},1)} \left\|\boldsymbol{G}\left(w_\beta^{\text{asy}}\right) - \boldsymbol{\nabla} w_\beta^{\text{asy}}\right\|_{L_2(K)}^2 . \tag{7.116}$$

The relationship (7.107) reveals that the estimators on the physical and reference cells are related by the formula

$$\epsilon\left((q_\beta)_X^{\text{asy}}, -\Delta q, \mathcal{C}(\boldsymbol{c}_\alpha, h)\right)^2 = h^4 \epsilon\left(w_\beta^{\text{asy}}, -\Delta w_\beta, \mathcal{C}(\boldsymbol{0}, 1)\right)^2 ; \tag{7.117}$$

therefore, it is sufficient to perform the computations over the reference cell.

The above manipulations would allow the asymptotic effectivity index to be computed for that particular case. Suppose, more generally, that the true solution u and the finite element mesh are chosen to satisfy conditions $(A1)$ and $(A2)$ but are otherwise arbitrary. The quadratic Taylor polynomial approximation to u is of the form

$$q = \sum_{|\beta| \le 2} a_\beta q_\beta \tag{7.118}$$

for suitable coefficients a_β, and, as before, the asymptotic finite element approximation to q is given by

$$q_X^{\text{asy}} = \sum_{|\beta|\le 2} a_\beta (q_\beta)_X^{\text{asy}} \tag{7.119}$$

or, in terms of the approximation on the reference cell,

$$q_X^{\text{asy}} = h^2 w^{\text{asy}} = h^2 \sum_{|\beta|\le 2} a_\beta w_\beta^{\text{asy}}. \tag{7.120}$$

The *a posteriori* error estimator may, in principle, be applied to the data q_X^{asy} and $-\Delta q$, yielding the estimator

$$\epsilon(q_X^{\text{asy}}, -\Delta q, \mathcal{C}(\boldsymbol{c}_\alpha, h))^2 = \sum_{K\subset\mathcal{C}(\boldsymbol{c}_\alpha,h)} \|\boldsymbol{G}\left(q_X^{\text{asy}}\right) - \boldsymbol{\nabla} q_X^{\text{asy}}\|^2_{L_2(K)}; \tag{7.121}$$

alternatively, the estimator can be written in terms of the estimator obtained using the approximation over the reference cell as in (7.120),

$$\epsilon(q_X^{\text{asy}}, -\Delta q, \mathcal{C}(\boldsymbol{c}_\alpha, h))^2 = h^4 \sum_{K\subset\mathcal{C}(\boldsymbol{0},1)} \|\boldsymbol{G}\left(w^{\text{asy}}\right) - \boldsymbol{\nabla} w^{\text{asy}}\|^2_{L_2(K)}. \tag{7.122}$$

Now, by the linearity of the recovery operator $\boldsymbol{G}$, it follows that

$$\boldsymbol{G}\left(w^{\text{asy}}\right) - \boldsymbol{\nabla} w^{\text{asy}} = \sum_{|\beta|=2} a_\beta \left(\boldsymbol{G}\left(w_\beta^{\text{asy}}\right) - \boldsymbol{\nabla} w_\beta^{\text{asy}}\right) \tag{7.123}$$

since the terms with $|\beta| \le 1$ vanish identically thanks to the consistency condition of the recovery operator. Inserting this expression into (7.122) and expanding leads to the representation

$$\epsilon(q_X^{\text{asy}}, -\Delta q, \mathcal{C}(\boldsymbol{c}_\alpha, h))^2 = h^4 \sum_{|\beta|=2} \sum_{|\gamma|=2} a_\beta \mathcal{E}_{\beta\gamma}^{\text{rec}} a_\gamma \tag{7.124}$$

where

$$\mathcal{E}_{\beta\gamma}^{\text{rec}} = \sum_{K\subset\mathcal{C}(\boldsymbol{0},1)} \left(\boldsymbol{G}\left(w_\beta^{\text{asy}}\right) - \boldsymbol{\nabla} w_\beta^{\text{asy}}, \boldsymbol{G}\left(w_\gamma^{\text{asy}}\right) - \boldsymbol{\nabla} w_\gamma^{\text{asy}}\right)_{L_2(K)} \tag{7.125}$$

and $(\cdot,\cdot)_{L_2(K)}$ denotes the L_2 inner product over the element. For the particular case of the reference cell shown in Figure 7.5, it will be found that for the recovery-based estimator of Section 4.6, we have

$$\mathcal{E}^{\text{rec}} = \frac{4}{9}\begin{bmatrix} 4(2+113\delta^2/50) & 129\delta^2/20 & -4(1+13\delta^2/50) \\ 129\delta^2/20 & 3(1+387\delta^2/100) & 129\delta^2/20 \\ -4(1+13\delta^2/50) & 129\delta^2/20 & 4(2+113\delta^2/50) \end{bmatrix}. \tag{7.126}$$

The expression for $\mathcal{E}^{\text{rec}}$ can be compared with the expression (7.113) for the matrix $\mathcal{E}$ obtained for the true error. In particular, if the central node is unperturbed—that is, $\delta = 0$—then $\mathcal{E} = \mathcal{E}^{\text{rec}}$ and we conclude that in this case

$$|||q - q_X^{\text{asy}}|||_{\mathcal{C}(\boldsymbol{c}_\alpha, h)} = \epsilon(q_X^{\text{asy}}, -\Delta q, \mathcal{C}(\boldsymbol{c}_\alpha, h)); \tag{7.127}$$

consequently, the asymptotic value of the effectivity index is unity.

If the node is perturbed, so that $\delta \neq 0$, then the asymptotic effectivity index will depend on the particular form of the quadratic polynomial q, or, equivalently, on the data a_β, according to the formula

$$\left[\frac{\epsilon(q_X^{\text{asy}}, -\Delta q, \mathcal{C}(\boldsymbol{c}_\alpha, h))}{|||q - q_X^{\text{asy}}|||_{\mathcal{C}(\boldsymbol{c}_\alpha, h)}}\right]^2 = \frac{\boldsymbol{a}^\top \mathcal{E}^{\text{rec}} \boldsymbol{a}}{\boldsymbol{a}^\top \mathcal{E} \boldsymbol{a}} \tag{7.128}$$

where $\boldsymbol{a}$ is the vector composed of the coefficients a_β.

A key observation in the development of this chapter is that the extreme values of the quotient appearing in (7.128) may be obtained by evaluating the relative eigenvalues of the matrices $\mathcal{E}$ and $\mathcal{E}^{\text{rec}}$, by solving the generalized eigenvalue problem

$$\mathcal{E}^{\text{rec}} \boldsymbol{a} = \lambda \mathcal{E} \boldsymbol{a}. \tag{7.129}$$

In turn, the value of the asymptotic effectivity index for a general solution u satisfying (A1) is included in this range, since its quadratic Taylor polynomial approximation corresponds to a particular selection of the vector $\boldsymbol{a}$. The relative eigenvalues of the matrices $\mathcal{E}^{\text{rec}}$ and $\mathcal{E}$ for the reference cell shown in Figure 7.5 are found to be

$$1 + \frac{21}{25}\delta^2$$

and

$$\frac{84\delta^4 + 1087\delta^2 + 200 \pm 3\delta^2\sqrt{(1 + 2\delta^2)(991 + 392\delta^2)}}{200(1 + 5\delta^2)}$$

and a plot of these eigenvalues is shown in Figure 7.6. The plot shows that the performance of the estimator degenerates as the mesh becomes more distorted and the estimator will tend to overestimate the true error. The information shown in the figure also shows the dangers of drawing conclusions concerning the general performance of an estimator on the basis of particular benchmark problems. For instance, if the particular set of benchmark problems had solutions whose behavior was similar to the functions corresponding to the smallest eigenvalue, then one would have to conclude that the estimator performed well for all values of δ. Nevertheless, the true picture is rather different and contains features that might easily be missed even after careful selection of the benchmark problems.

Incidentally, while analytic expressions have been obtained for the extreme effectivity indices in terms of the variable δ, in practical computations one would generate the curves by performing a number of repeat runs of the

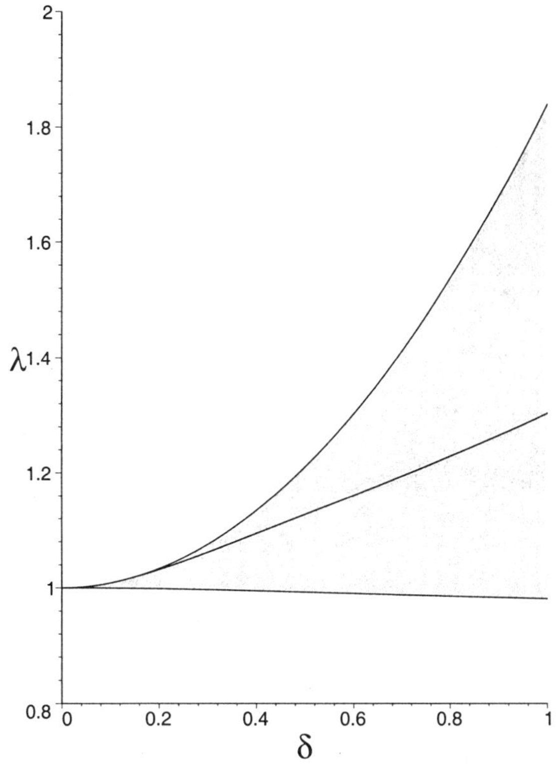

Fig. 7.6 Plot of the variation of relative eigenvalues of the matrices $\mathcal{E}^{\text{rec}}$ and $\mathcal{E}$ as the position (δ, δ) of the central node in the reference cell shown in Figure 7.5 is varied. For each particular value of δ, the square of the asymptotic effectivity for the recovery-based estimator applied to a general solution is contained between the extreme eigenvalues.

code with different values of δ. The purpose here was to provide analytic expressions that could be used to assist the reader in verifying their own code.

7.6.4 Application to Equilibrated Residual Method

An argument identical to the one pursued in the previous section for the recovery-based estimator may be applied to the equilibrated residual method of *a posteriori* error estimation. As before, the matrix of couplings between the estimators for each of the quadratic monomials is the main ingredient. A tedious computation reveals that for the equilibrated residual method, using fluxes constructed as in Chapter 6, in conjunction with quadratic bubble functions for the solution of the local residual problem, the matrix is given by

$$\mathcal{E}^{\mathsf{equ}} = \frac{4}{27(1-\delta^4)}\mathcal{E}_0^{\mathsf{equ}}$$

where $\mathcal{E}_0^{\mathsf{equ}}$ is the matrix given by

$$\begin{bmatrix} 23+14\delta^2-16\delta^4-12\delta^6 & 18\delta^2(1-\delta^4) & -13+14\delta^2+2\delta^4-12\delta^6 \\ 18\delta^2(1-\delta^4) & 9(1+3\delta^2)(1-\delta^4) & 18\delta^2(1-\delta^4) \\ -13+14\delta^2+2\delta^4-12\delta^6 & 18\delta^2(1-\delta^4) & 23+14\delta^2-16\delta^4-12\delta^6 \end{bmatrix}$$

for $\delta \neq 1$. The relative eigenvalues of this matrix compared with the matrix $\mathcal{E}$ formed using the true errors give bounds on the square of the asymptotic effectivity index of the method. These eigenvalues are found to be 1,

$$\frac{2-\delta^4}{2(1-\delta^4)} = 1 + \frac{1}{2}\delta^4 + \mathcal{O}(\delta^8)$$

and

$$\frac{5+34\delta^2+33\delta^4}{6+36\delta^2+30\delta^4} = \frac{5}{6} + \frac{2}{3}\delta^2 - \frac{8}{3}\delta^4 + \mathcal{O}(\delta^6)\,.$$

It will be observed that the largest of these eigenvalues blows up as the mesh becomes highly distorted in the limit $\delta \to 1$, meaning that the estimator may be excessively pessimistic on such meshes. This is shown clearly in Figure 7.7.

7.6.5 Application to Implicit Element Residual Method

The full dependence on δ of the matrix $\mathcal{E}^{\mathsf{imp}}$ and the relative eigenvalues for the estimator obtained using the implicit element residual method, as described in Bank and Weiser [37], is rather complicated and unilluminating. Therefore, only the leading terms in the series expansions are provided.

$$\mathcal{E}^{\mathsf{imp}} = \frac{4}{9}\begin{bmatrix} \frac{41}{3} & 0 & -\frac{31}{3} \\ 0 & 3 & 0 \\ -\frac{31}{3} & 0 & \frac{41}{3} \end{bmatrix} + \frac{4}{9}\begin{bmatrix} 17 & 18 & 17 \\ 18 & 21 & 18 \\ 17 & 18 & 17 \end{bmatrix}\delta^2 + \mathcal{O}(\delta^4)\,.$$

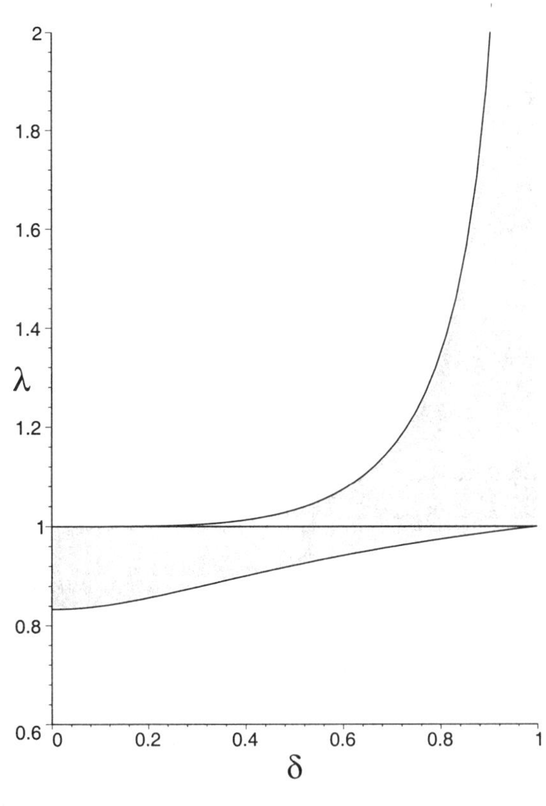

Fig. 7.7 Plot of the variation of relative eigenvalues of the matrices $\mathcal{E}^{\mathsf{equ}}$ and $\mathcal{E}$ as the position (δ, δ) of the central node in the reference cell shown in Figure 7.5 is varied. The largest eigenvalue becomes unbounded in the limit $\delta \to 1$.

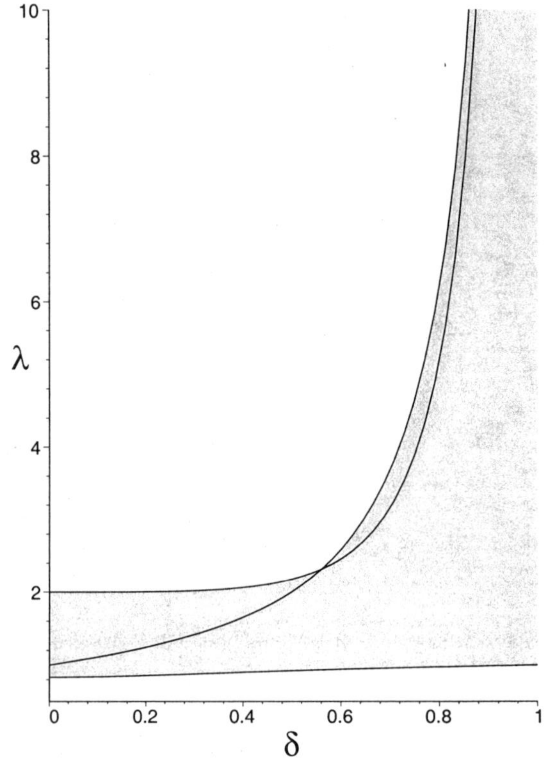

Fig. 7.8 Plot of the variation of relative eigenvalues of the matrices $\mathcal{E}^{\mathsf{imp}}$ and $\mathcal{E}$ as the position (δ, δ) of the central node in the reference cell shown in Figure 7.5 is varied. As $\delta \to 1$, the two largest eigenvalues become unbounded.

The leading terms of the series expansions for the eigenvalues are found to be

$$2 + 2\delta^4 + \mathcal{O}(\delta^8),$$

$$\frac{5}{6} + \frac{41}{6}\delta^2 - \frac{548}{3}\delta^4 + \mathcal{O}(\delta^6)$$

and

$$1 + 4\delta^2 + 132\delta^4 + \mathcal{O}(\delta^6),$$

while the exact eigenvalues are plotted in Figure 7.8. The two largest eigenvalues blow up as $\delta \to 1$, showing that the performance of the estimator is again extremely poor on such distorted meshes. Interestingly, even on the regular mesh where the central node is unperturbed, $\delta = 0$, the asymptotic effectivity index of the estimator could be as high as $\sqrt{2}$.

7.7 BIBLIOGRAPHICAL REMARKS

The material contained in this chapter is inspired by Babuška *et al.* [29], where the performance of estimators for piecewise affine finite approximation on triangular elements is considered. Extensive numerical results are presented where the performance of the estimators is studied with respect to the variation of the topology of the mesh cells, the anisotropy of the underlying operator, and the closely related issue of mesh anisotropy. The methodology presented here may be extended to incorporate effects due to the domain boundaries and the important question of the selection of boundary conditions for the estimators [31].

The notion of the asymptotic and periodic finite element approximation proves fruitful in the study of superconvergence effects in finite element analysis [26, 28]. The proximity of the periodic finite element approximation to the asymptotic finite element approximation means that the superconvergence points for the asymptotic finite element approximation coincide with those of the periodic finite element approximation. These may be identified by examining the periodic finite element approximation computed in the same fashion as described in the present chapter.

8

Estimation of the Errors in Quantities of Interest

8.1 INTRODUCTION

The goal of many finite element computations is the determination of a key, specific quantity $Q(u)$, depending on the solution u, that is needed for a particular design decision. One example of such a quantity would be the flux flowing out over a portion Γ of the boundary of the domain

$$Q(v) = \int_{\Gamma} \boldsymbol{n} \cdot \boldsymbol{q}(v) \, \mathrm{d}s$$

where $\boldsymbol{q}(v) = -a\boldsymbol{\nabla} v$ is the flux vector and $\boldsymbol{n}$ is the unit, outward normal vector. Alternatively, one might wish to estimate the value of the solution at a particular point $\boldsymbol{x}_0$,

$$Q(v) = v(\boldsymbol{x}_0).$$

The actual pointwise values of the solution and its gradient predicted by a particular model may well be "infinite" in critical regions of the domain such as re-entrant corners. Obviously, it is meaningless to attempt to estimate the value in such cases. Instead, one may be content to work with the average value of the quantity in a small neighborhood $\omega_\epsilon(\boldsymbol{x}_0)$ of the critical point, and define the quantity of interest to be

$$Q(v) = \frac{1}{|\omega_\epsilon(\boldsymbol{x}_0)|} \int_{\omega_\epsilon(\boldsymbol{x}_0)} v \, \mathrm{d}\boldsymbol{x}.$$

Often, the quantity of interest will be a vector-valued, or even tensor-valued, function, such as the average value of the stresses in some critical portion ω

of a structure

$$\frac{1}{|\omega|}\int_\omega \boldsymbol{\sigma}(\boldsymbol{u})\ \mathrm{d}\boldsymbol{x}$$

where $\boldsymbol{u}$ is the displacement of the (linearly elastic) structure and $\boldsymbol{\sigma}$ is the associated stress field. A set of linear functionals associated with this quantity of interest could be defined to be

$$Q_{ij}(\boldsymbol{v}) = \frac{1}{|\omega|}\int_\omega \sigma_{ij}(\boldsymbol{v})\ \mathrm{d}\boldsymbol{x},$$

with each functional dealt with separately in turn.

The estimation of the error e in the primary variable measured in the global energy norm has been the target of most of the analysis up to this point. Our purpose in the present chapter is to show how estimators for the error measured in the energy norm may be utilized for the recovery and error estimation of particular quantities of interest. In order to develop a reasonably general framework, it is convenient to regard a quantity of interest as a bounded, linear functional $Q : V \to \mathbb{R}$ acting on the space V of admissible functions for the problem at hand. The quantity of interest in each of the above examples may be represented in the form of a *linear* functional Q. However, as remarked above, for certain quantities of interest, such as pointwise values, the functional is unbounded and, strictly speaking, it is meaningless to attempt to estimate the quantity. Nevertheless, it is possible to treat such instances as bounded, linear functionals indirectly through the use of a closely related and *bounded* linear functional such as a local average, as suggested above, or by the use of appropriate extraction formulae.

The most straightforward way to estimate a quantity of interest Q in the true solution u given the finite element approximation u_X, is to simply take $Q(u) \approx Q(u_X)$. Is it possible to *quantify* the accuracy of this approximation? The linearity of the functional Q means that the error in the approximation may be written in the form

$$Q(u) - Q(u_X) = Q(e)$$

where $e = u - u_X$ is the error in the finite element approximation. This will form our point of departure in investigating to what extent the earlier work on estimating the error e measured in the *energy norm* may be exploited to obtain *quantitative estimates for the error in quantities of interest.* A systematic approach will be presented for calculating estimates of error in quantities of interest that may be characterized by bounded linear functionals Q. A notable feature of the theory to be presented is that it will be possible to obtain both *upper* and *lower* bounds for the errors in the quantity of interest. Thus, if $Q(e)$ is the error in the approximation of the quantity of interest $Q(u)$, then it will be shown how the techniques developed for estimating the error in the global energy norm may be used to compute quantities η^Q_{upp} and η^Q_{low} satisfying

$$\eta^Q_{\mathrm{low}} \le Q(e) \le \eta^Q_{\mathrm{upp}}.$$

An immediate consequence of these bounds for the error is that upper and lower bounds may be obtained for the quantity of interest itself,

$$Q(u_X) + \eta^Q_{\text{low}} \le Q(u) \le Q(u_X) + \eta^Q_{\text{upp}}.$$

The chapter concludes with an introductory discussion of pollution of local error estimates and on techniques to characterize and estimate pollution errors.

8.2 ESTIMATES FOR THE ERROR IN QUANTITIES OF INTEREST

Consider the model problem

$$u \in V : B(u, v) = L(v) \quad \forall v \in V$$

and its finite element approximation

$$u_X \in X : B(u_X, v_X) = L(v_X) \quad \forall v_X \in X$$

where $B(\cdot,\cdot)$ and $L(\cdot)$ are defined in Section 1.4. Let Q be a bounded linear functional representing some quantity of interest. The goal is to estimate the error in the approximation of the quantity of interest

$$Q(e) = Q(u) - Q(u_X) \tag{8.1}$$

with the aim of utilising estimators for the error measured in the energy norm. *Is there really any reason to suspect this goal is realistic?*

If the norm for the space V is chosen to be the energy norm, then the norm of Q is defined by

$$||Q||_{V'} = \sup_{v \in V} \frac{|Q(v)|}{\|\!|v|\!\|}. \tag{8.2}$$

This immediately implies the following bound for the error in the quantity of interest in terms of the energy norm of the error

$$|Q(e)| \le ||Q||_{V'} \, \|\!|e|\!\| \,. \tag{8.3}$$

In this fashion, it is possible to relate the error in the quantity of interest to the error in the finite element approximation measured in the energy norm.

Of course, the bound (8.3) is of limited use as it stands in several respects, not least of which is that the norm of the functional Q is unknown. Fortunately, there is a standard mathematical procedure for evaluating the norm of a bounded linear functional that applies to the present setting, based on the Riesz Representation Theorem. The *(Riesz) representor*, w^Q, of the bounded linear functional Q is defined to be the unique solution of the problem

$$w^Q \in V : B(v, w^Q) = Q(v) \quad \forall v \in V. \tag{8.4}$$

At this stage, we are assuming that the bilinear form B is symmetric, so there is no difference in placing the unknown w^Q in the second argument of B. However, it is essential for w^Q to appear in this fashion when unsymmetric problems are considered. With the aid of (8.2), it follows that

$$\|Q\|_{V'} = \sup_{v \in V} \frac{|B(v, w^Q)|}{|||v|||}$$

and then, by the Cauchy–Schwarz inequality, we obtain

$$|B(v, w^Q)| \leq |||w^Q||| \; |||v|||$$

and it follows that

$$\|Q\|_{V'} \leq |||w^Q||| .$$

This bound actually holds as an *equality* as is shown by choosing $v = w^Q$ in (8.4), so that

$$Q(w^Q) = |||w^Q|||^2$$

and hence

$$\|Q\|_{V'} \geq \frac{|Q(w^Q)|}{|||w^Q|||} = |||w^Q||| .$$

These considerations show that

$$|Q(e)| \leq |||w^Q||| \; |||e||| ; \tag{8.5}$$

consequently, it is entirely reasonable to expect to be able to obtain computable bounds in terms of the energy norm of the error.

The bound (8.5) suffers from two main defects as it stands. Firstly, it requires knowledge of the exact representor of Q. Secondly, the general argument leading to (8.5) takes no advantage of the fact that e is the error in a Galerkin approximation; as a result, it will transpire that the bound is rather pessimistic. Finally, it is unnecessary for the bilinear form to be symmetric. These shortcomings are removed in the following result:

Theorem 8.1 *Let $u_X \in X$ denote the finite element approximation of the solution of problem*

$$u \in V : B(u, v) = L(v) \quad \forall v \in V$$

where $B : V \times V \to \mathbb{R}$ is a continuous, coercive and bilinear form and $L : V \to \mathbb{R}$ is bounded and linear. Let $Q : V \to \mathbb{R}$ be a bounded, linear functional and define w^Q to be the solution of the adjoint problem,

$$w^Q \in V : B(v, w^Q) = Q(v) \quad \forall v \in V. \tag{8.6}$$

Then, the error in estimating the quantity $Q(u)$ using the finite element approximation $Q(u_X)$ is given by

$$Q(u) - Q(u_X) = B(e, e^Q) \tag{8.7}$$

where e and e^Q are the errors in the finite element approximation of u and w^Q, respectively.

Proof. The choice of $v = e$ in (8.6) shows that

$$Q(e) = B(e, w^Q).$$

The Galerkin orthogonality property implies that for all $w_X \in X$, we have

$$Q(e) = B(e, w^Q) = B(e, w^Q - w_X).$$

Selecting w_X to be finite element approximation of w^Q defined by the solution of the problem

$$w_X \in X : B(v_X, w_X) = Q(v_X) \quad \forall v_X \in X$$

then reveals

$$Q(e) = B(e, e^Q)$$

as claimed. ∎

The estimate (8.7) shows that the rate of convergence of the error in the quantity of interest will, roughly speaking, be *double* the rate of convergence of the error measured in the energy norm. In addition, it shows that the error in the quantity of interest Q may be bounded by the errors in the finite element approximation of the solution u and the representor of Q, *measured in the energy norm.*

It follows that any of the earlier estimators for the error measured in the energy norm may be invoked to estimate the error in the quantity of interest. Thus, the conclusions of Theorem 8.1 justify the considerable attention paid earlier to the estimation of error measured in the energy norm.

8.3 UPPER AND LOWER BOUNDS ON THE ERRORS

Suppose that η is an *a posteriori* estimator for the error e in the original problem, measured in the energy norm. In a similar fashion, given the finite element approximation to the Riesz representor, w^Q, of the quantity of interest Q satisfying

$$w_X^Q \in X : B(w_X^Q, v_X) = Q(v_X) \quad \forall v_X \in X, \tag{8.8}$$

one could use the same *a posteriori* estimation technique to produce a computable estimator for the error $e^Q = w^Q - w_X^Q$:

$$|||e^Q||| \leq \eta^Q. \tag{8.9}$$

Theorem 8.1 then shows that

$$|Q(u) - Q(u_X)| \leq \eta\eta^Q \tag{8.10}$$

and, consequently,

$$Q(u_X) - \eta\eta^Q \leq Q(u) \leq Q(u_X) + \eta\eta^Q. \tag{8.11}$$

In principle, these bounds represent the solution of the problem of estimating the error in quantities of interest. However, these bounds may be refined in certain circumstances, and in particular if the bilinear form B is *symmetric*.

The Cauchy–Schwarz inequality used to deduce the bound (8.10) from the *exact* formula (8.7) generally produces rather a pessimistic bound. In search of a means to sidestep the use of the Cauchy–Schwarz inequality, one may instead take advantage of the *parallelogram identity*, provided that the bilinear form B is symmetric:

$$B(e, e^Q) = \frac{1}{4}\left|\!\left|\!\left| \alpha e + \frac{1}{\alpha}e^Q \right|\!\right|\!\right|^2 - \frac{1}{4}\left|\!\left|\!\left| \alpha e - \frac{1}{\alpha}e^Q \right|\!\right|\!\right|^2, \quad \alpha \neq 0. \tag{8.12}$$

Applying this in conjunction with Theorem 8.1 provides an *exact* representation of the error in the quantity of interest:

$$Q(u) - Q(u_X) = Q(e) = \frac{1}{4}\left|\!\left|\!\left| \alpha e + \frac{1}{\alpha}e^Q \right|\!\right|\!\right|^2 - \frac{1}{4}\left|\!\left|\!\left| \alpha e - \frac{1}{\alpha}e^Q \right|\!\right|\!\right|^2. \tag{8.13}$$

This relationship provides the starting point for the derivation of upper and lower bounds on $Q(e)$.

The following simple result shows that the problem reduces to the derivation of upper and lower bounds on errors measured in the energy norm.

Lemma 8.2 *Let α be a fixed, nonzero real constant and suppose that η^+_{low}, η^+_{upp}, η^-_{low}, and η^-_{upp} satisfy*

$$\eta^+_{\text{low}} \leq |\!|\!| \alpha e + \alpha^{-1} e^Q |\!|\!| \leq \eta^+_{\text{upp}} \tag{8.14}$$

and

$$\eta^-_{\text{low}} \leq |\!|\!| \alpha e - \alpha^{-1} e^Q |\!|\!| \leq \eta^-_{\text{upp}}. \tag{8.15}$$

Then,

$$\frac{1}{4}\left(\eta^+_{\text{low}}\right)^2 - \frac{1}{4}\left(\eta^-_{\text{upp}}\right)^2 \leq Q(u) - Q(u_X) \leq \frac{1}{4}\left(\eta^+_{\text{upp}}\right)^2 - \frac{1}{4}\left(\eta^-_{\text{low}}\right)^2. \tag{8.16}$$

Proof. The results are a trivial consequence of (8.13). ■

Upper Bounds: Upper bounds for the errors in energy may be developed based on the equilibrated residual method of Chapter 6. Invoking Theorem 6.1, it follows that the residual for the error in the finite element approximation of u may be decomposed in the form

$$B(e, v) = L(v) - B(u_X, v) = \sum_{K\in\mathcal{P}} B_K(\phi_K, v), \quad \forall v \in V$$

where $\{\phi_K : K \in \mathcal{P}\}$ are the solutions of local residual problems (6.14). Equally well, it is possible to decompose the residual in the approximation of the representor into contributions from each element in the same way:

$$B(e^Q, v) = L(v) - B(v, w_X) = \sum_{K\in\mathcal{P}} B_K(\phi_K^Q, v), \quad \forall v \in V$$

where $\{\phi_K^Q : K \in \mathcal{P}\}$ are solutions of local residual problems similar to (6.14).

A direct consequence of these decompositions is that

$$B(\alpha e \pm \alpha^{-1} e^Q, v) = \sum_{K\in\mathcal{P}} B_K(\alpha\phi_K \pm \alpha^{-1}\phi_K^Q, v) \quad \forall v \in V.$$

This type of decomposition leads to a natural upper bound on the error in the energy norm as shown in Theorem 6.1. Similarly, applying the Cauchy–Schwarz inequality, we obtain

$$\left|B(\alpha e \pm \alpha^{-1} e^Q, v)\right| \le \sum_{K\in\mathcal{P}} \left\|\!\left|\alpha\phi_K \pm \alpha^{-1}\phi_K^Q\right|\!\right\|_K \||v|\|_K$$

and then, again by the Cauchy–Schwarz inequality, we have

$$\left|B(\alpha e \pm \alpha^{-1} e^Q, v)\right| \le \left\{\sum_{K\in\mathcal{P}} \left\|\!\left|\alpha\phi_K \pm \alpha^{-1}\phi_K^Q\right|\!\right\|_K^2\right\}^{1/2} \||v|\| .$$

The characterization of the energy norm through duality,

$$\left\|\!\left|\alpha e \pm \alpha^{-1} e^Q\right|\!\right\| = \sup_{v\in V} \frac{\left|B(\alpha e \pm \alpha^{-1} e^Q, v)\right|}{\||v|\|}, \tag{8.17}$$

at once yields the desired upper bounds,

$$\left\|\!\left|\alpha e \pm \alpha^{-1} e^Q\right|\!\right\|^2 \le \sum_{K\in\mathcal{P}} \left\|\!\left|\alpha\phi_K \pm \alpha^{-1}\phi_K^Q\right|\!\right\|_K^2 . \tag{8.18}$$

Lower Bounds: The characterization (8.17) of the energy norm using duality lends itself to the derivation of lower bounds on the error. In particular, (8.17) shows that for any nonzero function $\xi \in V$, the error is bounded below by

$$\left\|\!\left|\alpha e \pm \alpha^{-1} e^Q\right|\!\right\| \ge \frac{\left|B(\alpha e \pm \alpha^{-1} e^Q, \xi)\right|}{\||\xi|\|}.$$

If the resulting lower bound is to be sharp then care must be exercised in the selection of the function ξ. Perhaps the most convenient method involves simply selecting v from a subspace Y consisting of functions that vanish on the element boundaries,

$$Y = \bigotimes_{K\in\mathcal{P}} Y_K \subset \bigotimes_{K\in\mathcal{P}} H_0^1(K) \subset V. \tag{8.19}$$

Spaces of this type consisting of bubble functions have been described, for example, in Chapter 5. The function $\xi \in Y$ is defined in terms of the energy projection of the true errors e and e^Q onto the subspace Y:

$$\xi = \alpha P_Y e \pm \frac{1}{\alpha} P_Y e^Q. \tag{8.20}$$

The projection $P_Y e$ itself is readily constructed via the solution of an error residual-type problem posed over the subspace Y,

$$P_Y e \in Y : B(P_Y e, v) = B(e, v) = L(v) - B(u_X, v) \quad \forall v \in Y. \tag{8.21}$$

The property (8.19) means that this problem decouples into independent local problems posed over each of the elements

$$P_Y e = \sum_{K \in \mathcal{P}} \overline{\phi}_K \tag{8.22}$$

where

$$\overline{\phi}_K \in Y_K : B_K(\overline{\phi}_K, v) = (f, v)_K - B_K(u_X, v) \quad \forall v \in Y_K. \tag{8.23}$$

This problem is closely related to the residual problem (6.14) defining the local error indicator ϕ_K. Specifically, if the residual problem (6.14) were to be approximated by replacing the space V_K by the subspace Y_K, then (observing that the contributions in (6.14) from the boundary fluxes g_K vanishes thanks to the property $Y_K \subset H_0^1(K)$), one would arrive at precisely the local problem (8.23). Advantage may be taken of this fact when the functions ϕ_K and $\overline{\phi}_K$ are computed. The same observations hold equally well for the solutions ϕ_K^Q and $\overline{\phi}_K^Q$. The lower bounds obtained in this way are often surprisingly good. More elaborate methods for producing enhanced lower bounds have been proposed in references [25] and [91].

Selection of the Parameter α: Finally, it remains to select a suitable nonzero value for the parameter α appearing in the parallelogram identity. A practical criterion for determining α is to minimize the upper bounds:

$$\left|\left|\left| \alpha e \pm \frac{1}{\alpha} e^Q \right|\right|\right|_K^2 \le \sum_{K \in \mathcal{P}} \left|\left|\left| \alpha \phi_K \pm \frac{1}{\alpha} \phi_K^Q \right|\right|\right|_K^2 .$$

The same value minimizes both of these bounds and is given by

$$\alpha^4 = \frac{\sum_{K \in \mathcal{P}} |||\phi_K^Q|||_K^2}{\sum_{K \in \mathcal{P}} |||\phi_K|||_K^2}, \tag{8.24}$$

provided that the denominator is nonzero. Obviously, if the denominator vanishes, then both the estimated and true errors vanish, and the selection of α is a moot point.

8.4 GOAL-ORIENTED ADAPTIVE REFINEMENT

Traditionally, adaptive refinement algorithms have been founded on control of the error measured in the energy norm. Of course, controlling the global error has a beneficial effect on the accuracy of local errors and errors in quantities of interest. Nevertheless, if one is concerned only with certain specific quantities, then the catch-all approach of adaptivity based on global error in energy might easily result in superfluous and extensive adaptive resolution of features of the solution that have little bearing on the accuracy of the particular quantity of interest.

The availability of computable estimates for errors in the specific quantity of interest makes it possible to construct adaptive refinement schemes designed to target the error in the quantity of interest directly. Goal-oriented adaptivity frequently results in a dramatic reduction in the computational costs, since only the features of the solution that strongly affect the quantity of interest are resolved.

A simple goal-oriented adaptive refinement algorithm designed to produce an approximation of the quantity of interest Q to a relative error tolerance γ_{tol} would be as follows:

1. Construct an initial partition of $\mathcal{P}$ of the domain Ω and an initial finite element subspace X.

2. Compute the finite element approximation u_X from the space X and approximate the quantity of interest by $Q(u_X)$.

3. Compute upper and lower bounds on the error in the quantity of interest using the techniques described in the previous section.

4. If the accuracy fails to meet the tolerance γ_{tol}, then adaptively enrich the finite element subspace X as described below and return to Step 2. Otherwise, the computation is terminated and the estimate for the quantity of interest is output along with the upper and lower bounds on the error.

Adaptive enrichment is performed whenever the estimator for the relative error exceeds the tolerance γ_{tol}:

$$e_{\text{rel}} = \frac{|Q(e)|}{|Q(u)|} \approx \frac{|\epsilon^Q|}{|Q(u_X)|} \geq \gamma_{\text{tol}} \tag{8.25}$$

where ϵ^Q is an estimate of the error $Q(e)$ in the quantity of interest. As is customary, the global estimate ϵ^Q is broken into contributions from the individual elements:

$$\epsilon^Q = \sum_K \epsilon_K^Q \approx \sum_K \eta_K \eta_K^Q \tag{8.26}$$

where η_K and η_K^Q are the respective local indicators for the errors e and e^Q measured in the energy norm over element K. A popular adaptive refinement criterion involves earmarking element K for refinement whenever the indicated local error is within, say, 70% of the largest indicated error:

$$\epsilon_K^Q \geq 30\% \max_{K'} \epsilon_{K'}^Q. \tag{8.27}$$

8.5 EXAMPLE OF GOAL-ORIENTED ADAPTIVITY

The advantages of goal-oriented adaptive algorithm will be illustrated by presenting the numerical results obtained for the simple model problem

$$-\Delta u = f \text{ in } \Omega; \quad u = 0 \text{ on } \Gamma_D; \quad \frac{\partial u}{\partial n} = 0 \text{ on } \Gamma_N \tag{8.28}$$

where the domain Ω is the unit square $(0,1) \times (0,1)$, with Dirichlet conditions applied on the edge $\Gamma_D = \{(x,y) : 0 < x < 1,\ y = 0\}$ and Neumann conditions prescribed on the remainder of the boundary. The source term f is chosen so that the exact solution of the problem is

$$u(x,y) = 5x^2(1-x)^2(e^{10x^2} - 1)y^2(1-y)^2(e^{10y^2} - 1). \tag{8.29}$$

The solution is smooth, but possesses steep gradients in the neighborhood of the boundaries as shown in Figure 8.1. Despite the true solution being symmetric with respect to the line $x = y$, the numerical approximation does not necessarily share this property thanks to the unsymmetrical boundary conditions.

The initial finite element approximation was obtained using a piecewise bilinear subspace constructed on a uniform initial partition consisting of 64 square elements.

8.5.1 Adaptivity Based on Control of Global Error in Energy

An adaptive refinement procedure of the type described in Section 8.4 was employed with the goal of controlling the global error measured in the energy norm. The enrichment of the finite element subspace was performed by subdividing those elements earmarked for refinement. The partition obtained after eight refinement passes is shown in the upper part of Figure 8.2. It will be observed that the steep gradients in the true solution have resulted in a highly refined partition in the neighborhood of the boundary.

8.5.2 Goal-Oriented Adaptivity Based on Pointwise Quantities of Interest

Suppose that we are required to produce an estimate for the value of the solution at a single point $\boldsymbol{x}_0$. Unfortunately, as pointed out earlier, in two or more dimensions, the pointwise values of the solution are not well-defined as

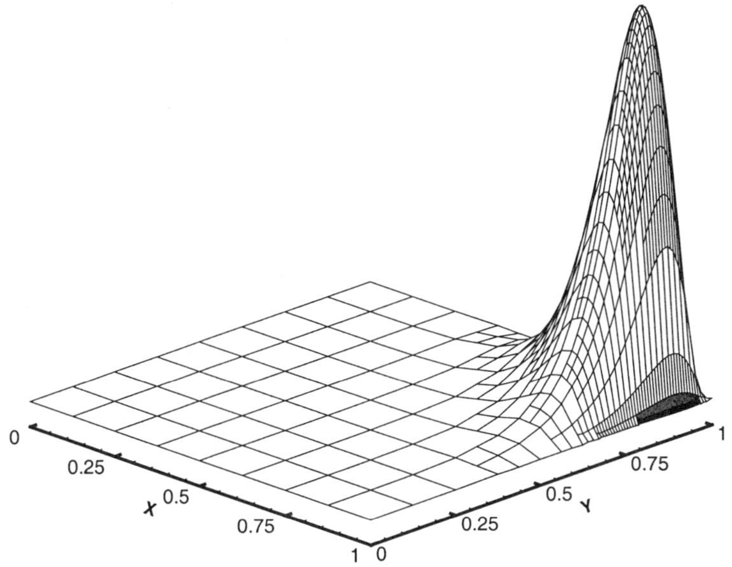

Fig. 8.1 The true solution of the model problem.

shown by the Sobolev Embedding Theorem. In general, pointwise evaluation functionals such as

$$Q(v) = v(\boldsymbol{x}_0)$$

are linear but not bounded on the space of admissible displacements. In order to circumvent this problem, one may instead work with a weighted local average, or *mollification*, of the solution u in the neighborhood of the point $\boldsymbol{x}_0$. The mollification of u is the smooth function u_ϵ defined, for given $\epsilon > 0$, by

$$u_\epsilon(\boldsymbol{x}_0) = \int_\Omega \rho_\epsilon(\boldsymbol{x}_0 - \boldsymbol{x}) u(\boldsymbol{x})\, \mathrm{d}\boldsymbol{x} \tag{8.30}$$

where the kernel $\rho_\epsilon \in C_0^\infty(\mathbb{R}^2)$ is nonzero on a ball of radius ϵ centered at $\boldsymbol{x}_0$. Typically, for points $\boldsymbol{x}_0$ located away from the boundary of the domain, ρ_ϵ is chosen to be the function

$$\rho_\epsilon(\boldsymbol{x}) = \begin{cases} c\exp[-1/(|\boldsymbol{x}|^2 - \epsilon^2)] & \text{if } |\boldsymbol{x}| < \epsilon \\ 0 & \text{otherwise} \end{cases}$$

where c is a constant that depends on ϵ and chosen such that the integral of ρ_ϵ is unity. It may be shown that if u is continuous at the point $\boldsymbol{x}_0$, then as the size ϵ of the ball shrinks, the values $u_\epsilon(\boldsymbol{x}_0)$ converge to the value of u at the point $\boldsymbol{x}_0$.

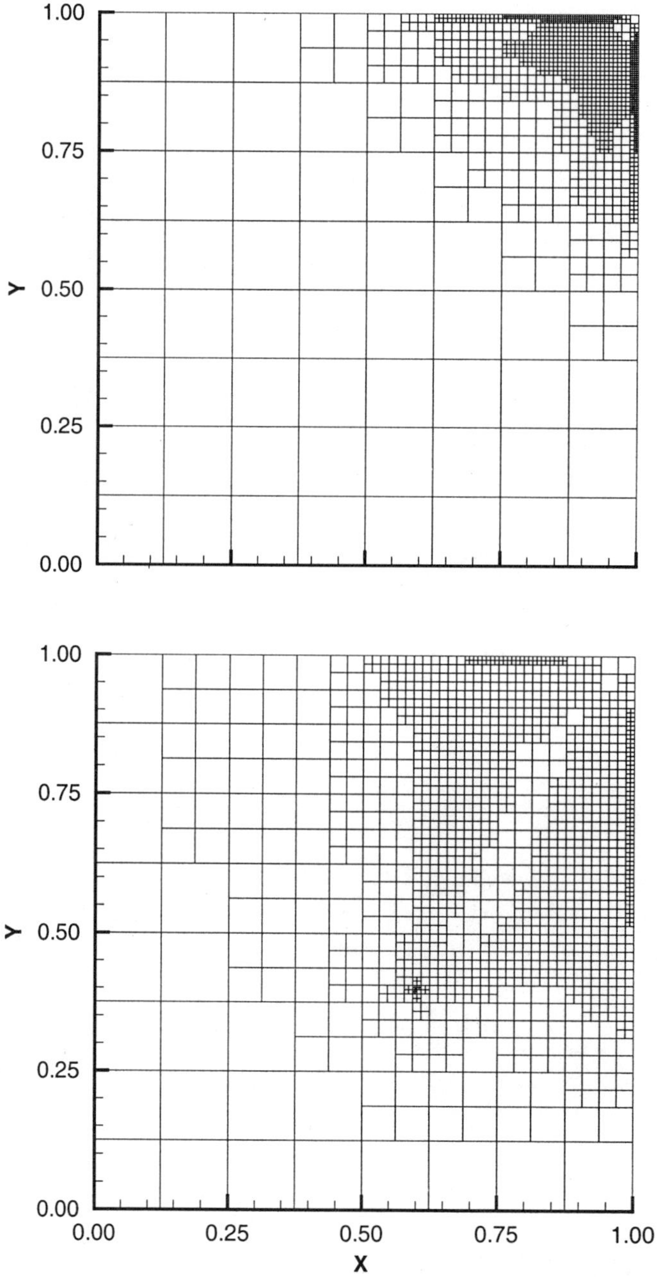

Fig. 8.2 Comparison of partitions generated by performing adaptive refinement based on controlling global error in energy norm (upper) versus goal-oriented adaptive analysis based on mollified pointwise value (lower). The target accuracy was $\gamma_{\mathrm{tol}} = 5.0\%$.

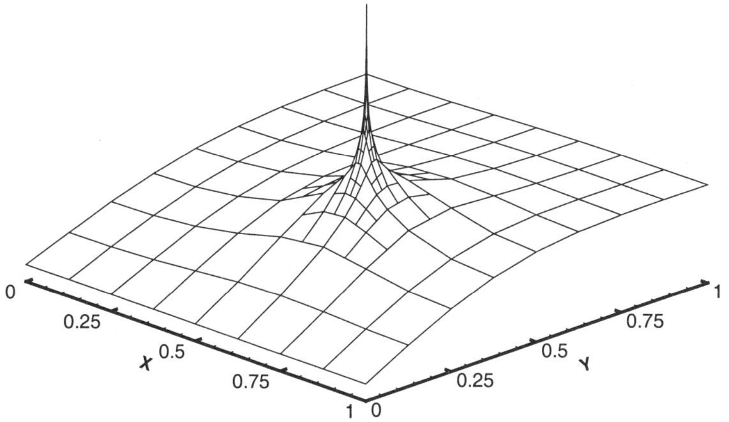

Fig. 8.3 Finite element approximation of the representor for the mollified pointwise value functional Q_ϵ in the case $\epsilon = 0.001$.

In view of these facts, the quantity of interest is defined for each given value of ϵ to be

$$Q_\epsilon(v) = v_\epsilon(\boldsymbol{x}_0). \tag{8.31}$$

For each fixed $\epsilon > 0$, this represents a bounded, linear functional on $H^1(\Omega)$. However, it should be borne in mind that the functional is unbounded in the limit $\epsilon \to 0$, and it is quite possible that the numerical bounds will deteriorate or even blow-up in this limit.

The use of the mollifier will be illustrated by performing goal-oriented adaptivity based on the desire to estimate the value of the solution u at the point $\boldsymbol{x}_0 = (0.6, 0.4)$. The value of ϵ is fixed as 10^{-3}. The numerical approximation of the representor associated with this quantity of interest is shown in Figure 8.3. The goal-oriented adaptive analysis is performed on the basis of the error estimates to a tolerance γ_{tol} of 5.0%. The final adapted mesh is shown the lower part of Figure 8.2 and consists of 1351 elements. The value of the quantity of interest on the final mesh is given by $Q_\epsilon(u_X) = 2.46$ whereas the true pointwise value is $u(\boldsymbol{x}_0) = 2.33$, giving a relative error of around 5.3%.

The advantages of goal-oriented adaptivity versus adaptivity based on control of the global energy norm are illustrated in Figure 8.4. It will be observed that the rate of convergence for the quantity of interest is significantly better when performing goal-oriented adaptivity.

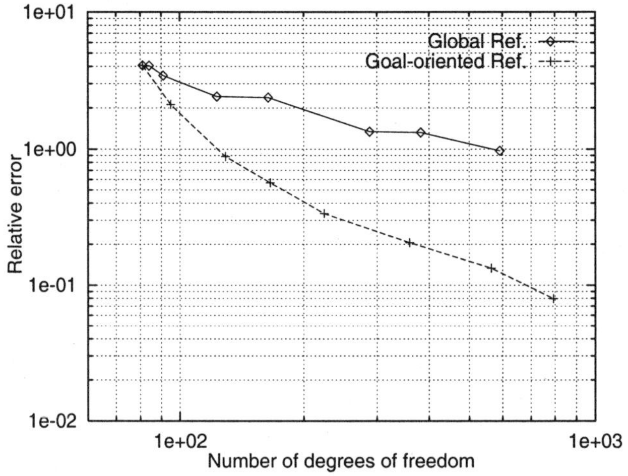

Fig. 8.4 Comparison of the relative error in the quantity of interest on meshes generated using control of global error in energy versus goal-oriented refinement.

8.6 LOCAL AND POLLUTION ERRORS

The advantage of goal-oriented adaptivity is that refinements are directed towards where their effect is most beneficial to the accuracy of the quantity of interest. However, even if the quantity of interest is based on purely local information, such as an average over a subdomain ω, then *it is generally insufficient to simply adapt the mesh solely on the basis of the error indicators over the subdomain* ω. As a matter of fact, the same conclusion holds for any superset of ω, unless it coincides with the entire domain. The justification of these claims relies on answering the following question. Let $\omega \subset \Omega$ be a subdomain comprising of one or more elements, and let η_K be an error indicator associated with element K. *What does the sum of the local indicators over* ω *really measure?*

Let ϕ_K, $K \in \mathcal{P}$, be the error indicator function generated through the equilibrated residual method, with associated local indicator $\eta_K = \|\phi_K\|_K$. The true error e is related to the indicator functions through the fundamental identity (6.21),

$$e \in V : B(e, v) = \sum_{K \in \mathcal{P}} B_K(\phi_K, v) \quad \forall v \in V. \tag{8.32}$$

A type of local error, e_ω^{loc}, over the subdomain ω may be defined by

$$e_\omega^{\text{loc}} \in V : B(e_\omega^{\text{loc}}, v) = \sum_{K \subset \omega} B_K(\phi_K, v) \quad \forall v \in V. \tag{8.33}$$

The local error indicator function ϕ_K depends only on the local residuals in the finite element approximation evaluated over the patch $\widetilde{K}$ of elements consisting of element K and its neighbors. In turn, this implies that the quantity e_ω^{loc} depends entirely on local residuals over the patch $\widetilde{\omega}$ consisting of the elements in ω and their neighboring elements. It is for this reason that we speak of e_ω^{loc} as a local error associated with the subdomain ω. The remaining component of the true error e is referred to as the nonlocal, or *pollution* error e_ω^{pol} and satisfies

$$e_\omega^{\text{pol}} \in V : B(e_\omega^{\text{pol}}, v) = \sum_{K \not\subset \omega} B_K(\phi_K, v) \quad \forall v \in V. \tag{8.34}$$

This part of the error emanates from residuals located outside the subdomain, yet are propagated into the subdomain thereby polluting the local accuracy within the subdomain.

Theorem 8.3 *Let $\omega \subset \Omega$ be a subdomain consisting of either a single element, a simply connected patch of elements or the entire domain Ω, and define the local error e_ω^{loc} as in (8.34), where ϕ_K is the local error indicator function associated with the equilibrated residual method. Then,*

$$|||e_\omega^{\text{loc}}|||^2 \le \sum_{K \subset \omega} |||\phi_K|||_K^2 \tag{8.35}$$

and there exists a positive constant C that is independent of the mesh size such that

$$\sum_{K \subset \omega} |||\phi_K|||_K^2 \le |||e_\omega^{\text{loc}}|||_{\widetilde{\omega}}^2 + C \sum_{K \subset \widetilde{\omega}} \left\{ h_K^2 \, \|\overline{r} - r\|_{L_2(K)}^2 + h_K \left\|\overline{R} - R\right\|_{L_2(\partial K)}^2 \right\} \tag{8.36}$$

where $\overline{r}$ and $\overline{R}$ are piecewise constant approximations to the interior r and interelement R residuals respectively.

Proof. An immediate consequence of (8.33) is that

$$\left|B(e_\omega^{\text{loc}}, v)\right| \le \sum_{K \in \omega} |||\phi_K|||_K \, |||v|||_K$$

and hence

$$\left|B(e_\omega^{\text{loc}}, v)\right| \le \left\{ \sum_{K \in \omega} |||\phi_K|||_K^2 \right\}^{1/2} |||v|||_\omega .$$

Consequently,

$$|||e_\omega^{\text{loc}}|||^2 \le \sum_{K \in \omega} |||\phi_K|||_K^2 . \tag{8.37}$$

As in Chapter 6, it may be shown that

$$|||\phi_K|||_K \le C \sum_{K' \subset \widetilde{K}} \left\{ h_K \, \|r\|_{L_2(K')} + h_K^{1/2} \, \|R\|_{L_2(\partial K')} \right\}$$

where r and R are the usual interior and interelement residuals. It was shown in Theorem 2.5 that the residuals could be bounded in terms of the error e evaluated over appropriate subdomains. The same method of proof used in Theorem 2.5 may be employed to obtain the bounds

$$\|r\|_{L_2(K)} \leq C\left\{h_K^{-1} \left\|\left|e_\omega^{\mathrm{loc}}\right|\right\|_K + \|\overline{r} - r\|_{L_2(K)}\right\} \tag{8.38}$$

and

$$\|R\|_{L_2(\gamma)} \leq C\left\{h_\gamma^{-1/2} \left\|\left|e_\omega^{\mathrm{loc}}\right|\right\|_{\widetilde{\gamma}} + h_\gamma^{1/2} \|\overline{r} - r\|_{L_2(\widetilde{\gamma})} + \left\|\overline{R} - R\right\|_{L_2(\gamma)}\right\}, \tag{8.39}$$

where the notation is defined in Chapter 2. These estimates at once imply that

$$\|\!|\phi_K|\!\|_K \leq C \left\|\left|e_\omega^{\mathrm{loc}}\right|\right\|_{\widetilde{K}} + C \sum_{K' \in \widetilde{K}} \left\{h_K \|\overline{r} - r\|_{L_2(K')} + h_K^{1/2} \left\|\overline{R} - R\right\|_{L_2(\partial K')}\right\} \tag{8.40}$$

and the result follows as claimed. ∎

This result shows that *the sum of the indicators over a subdomain does not measure the error over the subdomain.* Instead, the indicators measure the local error e_ω^{loc} generated by the local residuals. Pollution effects due to the remote residuals outside the subdomain are not visible to the local estimators. On reflection, this conclusion is only what is to be expected since the local indicator functions are completely independent of the residuals outside the subdomain. In fact, the conclusion of Theorem 8.3 holds (up to additional multiplicative constants C) for every estimator that is based on purely local information.

One implication is that if one were to adaptively refine a mesh using only the indicators over a subdomain ω, then it is *not* necessarily true that the error over the subdomain will be reduced. The local refinements reduce the local error e_ω^{loc} but have only an indirect effect on the pollution error e_ω^{pol}. Consequently, local refinements over the subdomain ω must be properly balanced with adequate refinement outside the subdomain to control the pollution error.

The identity (8.7) used in conjunction with (8.32) reveals that

$$Q(u) - Q(u_X) = B(e, e^Q) = \sum_{K \in \mathcal{P}} B(\phi_K, e^Q),$$

and hence

$$|Q(u) - Q(u_X)| \leq \sum_{K \in \mathcal{P}} \|\!|\phi_K|\!\|_K \left\|\left|e^Q\right|\right\|_K.$$

The goal-oriented adaptive algorithms presented in this chapter are based on this type of estimate. The proper balance between local and nonlocal refinements is maintained due to the presence of indicators from the entire mesh weighted by the error in the approximation of the representor function w^Q. The error in the approximation of the representor is typically larger in the neighborhood influencing the quantity of interest, thereby biasing the refinements towards the elements in the subdomain while properly taking account of the remote residuals.

8.7 BIBLIOGRAPHICAL REMARKS

The recovery of quantities of interest from finite element approximations has a long history. Techniques for extracting pointwise quantities were developed by Louis [82] along with error estimates showing the doubling of the rate of convergence in favorable circumstances. The series of articles of Babuška and Miller [16, 17, 18] proposed ideas similar to those presented in the text, including *a posteriori* error bounds. The widespread adoption of adaptive finite element techniques and *a posteriori* error estimation has been accompanied by a surge of interest in recovery of quantities of interest and goal-oriented adaptivity in general [47, 75, 95, 97, 98, 99, 101, 116]. The numerical examples presented in the text are taken from Oden and Prudhomme [92].

Issues related to the design and robustness of adaptive algorithms have been discussed by Dörfler [58, 59].

The discussion of local and pollution errors is based on the ideas presented in references [27] and [30]. Related work on local and pollution errors will be found in references [4, 71, 90].

9

Some Extensions

9.1 INTRODUCTION

Up to this point, we have developed theories and methods for *a posteriori* error estimation for a restricted class of model problems involving self-adjoint, linear elliptic operators with scalar valued solutions on planar domains. Remarkably, most of the results and techniques that have been developed thus far may be extended to significantly more general classes of problems. For the most part, such generalizations involve relatively straightforward extensions of the ideas developed in earlier chapters.

In the present chapter, we provide a brief account of how such extensions can be derived for important classes of problems, including indefinite problems, problems characterized by systems of coupled partial differential equations, problems with nonself-adjoint operators, and nonlinear partial differential equations. As models for problems with these features, we first consider the linear Oseen (and Stokes) equations, modeling viscous, incompressible flow in a given convective field. Here, we make use of saddle point theory and duality to derive implicit error estimators, including the equilibrated residual method. This paves the way to a treatment of the full Navier–Stokes equations for steady incompressible flow in the case of laminar flow. Finally, we consider a more abstract class of nonlinear elliptic problems and once again show how much of the methodologies dealt with earlier in this volume can be applied to obtain estimators.

9.2 STOKES AND OSEEN'S EQUATIONS

There are a number of specific issues that must be resolved when developing *a posteriori* error estimates for the Stokes problem. Firstly, the Stokes problem involves an incompressibility constraint and one must decide how to take proper account of the condition. Secondly, the Oseen approximation of the incompressible Navier–Stokes equations contains a nonself-adjoint term in the momentum part of the equations. This means that there is no natural energy norm in which to measure the error.

In developing implicit estimators for the Stokes problem, one might reasonably expect to be faced with an element residual problem requiring the solution of a local *Stokes* problem. However, there are drawbacks to this approach. For instance, it has been shown in earlier chapters that the local residual problem has to be approximated using an appropriate subspace. However, when dealing with the Stokes problem, the subspaces used to approximate the pressure and velocity components should also be constructed so that the *inf–sup* stability condition is satisfied. Consequently, one faces additional, rather awkward difficulties in designing stable schemes with which to approximate the local problem.

Firstly, the basic question of the norm in which the error will be estimated is considered. One outcome is that, perhaps surprisingly, it is unnecessary to solve a local Stokes problem in order to obtain an *a posteriori* error estimate. The significance of this conclusion in the design of a general *a posteriori* error estimation procedure for incompressible fluid flow is vital. In particular, the approximation of the local residual problems can, to a large extent, be developed independently of the type of element used to approximate the original fluid flow problem since there is no *inf–sup* condition to be satisfied. The analysis is valid for essentially any conforming discretization scheme for the Stokes problem. The approach reveals an appropriate equilibration principle for the determination of the boundary data, and the error estimator provides an upper bound on the true error in an energy like norm.

Let function spaces $\boldsymbol{V}$ and W be defined as follows

$$\boldsymbol{V} = H_0^1(\Omega) \times H_0^1(\Omega)$$

and

$$W = \left\{ q \in L_2(\Omega) : \int_\Omega q(\boldsymbol{x})\, \mathrm{d}\boldsymbol{x} = 0 \right\}.$$

Let $B : \boldsymbol{V} \times \boldsymbol{V} \to \mathbb{R}$ and $b : \boldsymbol{V} \times W \to \mathbb{R}$ be the bilinear forms

$$b(\boldsymbol{v}, q) = -\int_\Omega q \,\mathbf{div}\, \boldsymbol{v}\, \mathrm{d}\boldsymbol{x}$$

and

$$B(\boldsymbol{v}, \boldsymbol{w}) = \int_\Omega \{\nu \boldsymbol{\nabla} \boldsymbol{v} \cdot \boldsymbol{\nabla} \boldsymbol{w} + \boldsymbol{w} \cdot (\boldsymbol{U} \cdot \boldsymbol{\nabla})\, \boldsymbol{v}\}\, \mathrm{d}\boldsymbol{x}$$

where $\nu > 0$ is the viscosity parameter and $\boldsymbol{U}$ is a smooth solenoidal vector field on Ω (that is, $\mathbf{div}\,\boldsymbol{U} = 0$). For given data $\boldsymbol{f} \in L_2(\Omega) \times L_2(\Omega)$, we seek the solution of the problem:

Find $(\boldsymbol{u}, p) \in \boldsymbol{V} \times W$ such that for all $(\boldsymbol{v}, q) \in \boldsymbol{V} \times W$ we have

$$B(\boldsymbol{u}, \boldsymbol{v}) + b(\boldsymbol{v}, p) + b(\boldsymbol{u}, q) = F(\boldsymbol{v}) \tag{9.1}$$

where $F : \boldsymbol{V} \to \mathbb{R}$ is the linear functional

$$F(\boldsymbol{v}) = \int_\Omega \boldsymbol{f} \cdot \boldsymbol{v} \;\mathrm{d}\boldsymbol{x}. \tag{9.2}$$

Equation (9.1) may be written equivalently as a pair of equations by choosing $\boldsymbol{v} = \boldsymbol{0}$ and $q = 0$ in turn. For ease of exposition we consider only homogeneous Dirichlet boundary conditions. More general conditions may be dealt with in an analogous fashion. In order to describe sufficient conditions for the existence of a solution to (9.1) we introduce inner products $a(\cdot, \cdot)$ and $c(\cdot, \cdot)$ on $\boldsymbol{V}$ and W, respectively:

$$a(\boldsymbol{v}, \boldsymbol{w}) = \int_\Omega \nu \boldsymbol{\nabla} \boldsymbol{v} \cdot \boldsymbol{\nabla} \boldsymbol{w} \,\mathrm{d}\boldsymbol{x}$$

and

$$c(p, q) = \int_\Omega p\, q \;\mathrm{d}\boldsymbol{x}.$$

These inner products induce norms on $\boldsymbol{V}$ and W denoted by $||\cdot||_a$ and $||\cdot||_c$, respectively. The following facts [68] concerning B and b will be useful:

- There exists a positive constant C_B such that

$$|B(\boldsymbol{v}, \boldsymbol{w})| \le C_B \, ||\boldsymbol{v}||_a \;\; ||\boldsymbol{w}||_a \quad \forall \boldsymbol{v}, \; \boldsymbol{w} \in \boldsymbol{V}. \tag{9.3}$$

- The bilinear form b is continuous on $\boldsymbol{V} \times W$,

$$|b(\boldsymbol{v}, q)| \le \nu^{-1/2} \, ||\boldsymbol{v}||_a \;\; ||q||_c \quad \forall (\boldsymbol{v}, q) \in \boldsymbol{V} \times W. \tag{9.4}$$

- The *inf–sup* condition holds; that is, there exists a positive constant α_b such that

$$\sup_{\boldsymbol{v} \in \boldsymbol{V}} \frac{|b(\boldsymbol{v}, q)|}{||\boldsymbol{v}||_a} \ge \alpha_b \, ||q||_c \quad \forall q \in W. \tag{9.5}$$

- The bilinear form B is coercive; that is, owing to the vector field $\boldsymbol{U}$ being solenoidal there holds

$$B(\boldsymbol{v}, \boldsymbol{v}) = ||\boldsymbol{v}||_a^2 \quad \forall \boldsymbol{v} \in \boldsymbol{V}. \tag{9.6}$$

Under these conditions it follows [68] that there is a unique solution to (9.1).

The usual choice of norm on the product space $\boldsymbol{V} \times W$ is

$$(\boldsymbol{v}, q) \mapsto \left\{ \|\boldsymbol{v}\|_a^2 + \|q\|_c^2 \right\}^{1/2}.$$

It will be convenient to establish an equivalent norm for the space $\boldsymbol{V} \times W$. Let $(\boldsymbol{e}, E) \in \boldsymbol{V} \times W$ be *arbitrary*. Later, $\boldsymbol{e}$ and E will be chosen to be the error in the finite element approximation of the velocity $\boldsymbol{u}$ and pressure p, respectively. The pair $(\boldsymbol{\phi}, \psi) \in \boldsymbol{V} \times W$ is defined to be the Ritz projection of the residuals. That is,

$$a(\boldsymbol{\phi}, \boldsymbol{v}) + c(\psi, q) = B(\boldsymbol{e}, \boldsymbol{v}) + b(\boldsymbol{v}, E) + b(\boldsymbol{e}, q) \tag{9.7}$$

for all $(\boldsymbol{v}, q) \in \boldsymbol{V} \times W$. The existence and uniqueness of the pair $(\boldsymbol{\phi}, \psi)$ follows from the continuity of the forms B and b. Therefore, we may define

$$|\!|\!|(\boldsymbol{e}, E)|\!|\!| = \left\{ \|\boldsymbol{\phi}\|_a^2 + \|\psi\|_c^2 \right\}^{1/2}.$$

The following result confirms that this quantity is a norm on $\boldsymbol{V} \times W$ equivalent to the usual norm:

Theorem 9.1 *Under the foregoing assumptions and definitions, there exist positive constants k_1 and k_2 such that*

$$k_1 \, |\!|\!|(\boldsymbol{e}, E)|\!|\!|^2 \leq \|\boldsymbol{e}\|_a^2 + \|E\|_c^2 \leq k_2 \, |\!|\!|(\boldsymbol{e}, E)|\!|\!|^2$$

where k_1 depends only on C_B and ν, and k_2 depends only on C_B and α_b.

Proof. *Right Hand Inequality.* Making use of (9.5), (9.7), (9.3), and the Cauchy–Schwarz inequality yields

$$\|E\|_c \leq \frac{1}{\alpha_b} \left\{ \|\boldsymbol{\phi}\|_a + C_B \|\boldsymbol{e}\|_a \right\} \tag{9.8}$$

Using (9.6), (9.7) (with $q = -E$ and $\boldsymbol{v} = \boldsymbol{e}$), and the Cauchy–Schwarz inequality, we obtain

$$\|\boldsymbol{e}\|_a^2 \leq \|\boldsymbol{\phi}\|_a \|\boldsymbol{e}\|_a + \|\psi\|_c \|E\|_c \tag{9.9}$$

From (9.8) and (9.9) one finds

$$\frac{1}{2} \|\boldsymbol{e}\|_a^2 \leq \frac{1}{2} \left(\|\boldsymbol{\phi}\|_a + \frac{C_B}{\alpha_b} \|\psi\|_c \right)^2 + \frac{1}{\alpha_b} \|\boldsymbol{\phi}\|_a \|\psi\|_c \tag{9.10}$$

Combining (9.10) with (9.8), and once again using (9.10) yields the result with a constant k_2 depending on α_b and C_B.

Left Hand Inequality. Using (9.7), (9.3), (9.4), and the Cauchy–Schwarz inequality gives

$$\|\boldsymbol{\phi}\|_a \leq C_B \|\boldsymbol{e}\|_a + \nu^{-1/2} \|E\|_c \tag{9.11}$$

Using (9.7) and (9.4) gives

$$\|\psi\|_c^2 = b(\boldsymbol{e}, \psi) \leq \nu^{-1/2} \|\boldsymbol{e}\|_a \|\psi\|_c \tag{9.12}$$

Combining (9.11) and (9.12) yields the estimate claimed where k_1 depends only on C_B and ν. ■

Let $\mathcal{P}$ be a regular partitioning of the domain Ω. The finite element subspaces $\boldsymbol{X}$ and M are constructed in the usual manner so that the inclusion $\boldsymbol{X} \times M \subset \boldsymbol{V} \times W$ holds. The finite element approximation to (9.1) is then:

Find $(\boldsymbol{u}_X, p_M) \in \boldsymbol{X} \times M$ such that for all $(\boldsymbol{v}_X, q_M) \in \boldsymbol{X} \times M$

$$B(\boldsymbol{u}_X, \boldsymbol{v}_X) + b(\boldsymbol{v}_X, p_M) + b(\boldsymbol{u}_X, q_M) = F(\boldsymbol{v}_X)$$

A few remarks concerning the construction of the finite element subspace $\boldsymbol{X} \times M$ are in order. It will have been noted that there was no requirement for a discrete *inf–sup* condition to hold. The stability of the discretization scheme does not affect the *a posteriori* error analysis since only stability of the underlying continuous problem is used. Of course, the indiscriminate use of an unstable discretization scheme is not to be recommended.

9.2.1 A Posteriori Error Analysis

The local velocity space on each subdomain $K \in \mathcal{P}$ is

$$\boldsymbol{V}_K = \left\{\boldsymbol{v} \in H^1(K) \times H^1(K) : \boldsymbol{v} = \boldsymbol{0} \text{ on } \partial\Omega \cap \partial K\right\}$$

and the local pressure space is

$$W_K = L_2(K)$$

The bilinear forms $B_K : \boldsymbol{V}_K \times \boldsymbol{V}_K \to \mathbb{R}$ and $b_K : \boldsymbol{V}_K \times W_K \to \mathbb{R}$ are defined as follows:

$$b_K(\boldsymbol{v}, q) = -\int_K q \, \mathbf{div}\, \boldsymbol{v} \, \mathrm{d}x$$

and

$$B_K(\boldsymbol{v}, \boldsymbol{w}) = \int_K \{\nu \boldsymbol{\nabla} \boldsymbol{v} : \boldsymbol{\nabla} \boldsymbol{w} + \boldsymbol{w} \cdot (\boldsymbol{U} \cdot \boldsymbol{\nabla}) \boldsymbol{v}\} \, \mathrm{d}x.$$

Similarly, $F_K : \boldsymbol{V}_K \to \mathbb{R}$ is defined by

$$F_K(\boldsymbol{v}) = \int_K \boldsymbol{f} \cdot \boldsymbol{v} \, \mathrm{d}x$$

Hence for $\boldsymbol{v}$, $\boldsymbol{w} \in \boldsymbol{V}$ and $q \in W$ we have

$$b(\boldsymbol{v}, q) = \sum_{K \in \mathcal{P}} b_K(\boldsymbol{v}_K, q_K)$$

$$B(\boldsymbol{v}, \boldsymbol{w}) = \sum_{K \in \mathcal{P}} B_K(\boldsymbol{v}_K, \boldsymbol{w}_K)$$

and

$$F(\boldsymbol{v}) = \sum_{K \in \mathcal{P}} F_K(\boldsymbol{v}_K).$$

The broken velocity space $\boldsymbol{V}(\mathcal{P})$ is defined by

$$\boldsymbol{V}(\mathcal{P}) = \prod_{K \in \mathcal{P}} \boldsymbol{V}_K$$

and the broken pressure space $W(\mathcal{P})$ is defined by

$$W(\mathcal{P}) = \left\{ q \in \prod_{K \in \mathcal{P}} W_K : \int_\Omega q(\boldsymbol{x}) \, \mathrm{d}\boldsymbol{x} = 0 \right\}.$$

Examining the previous notations reveals that $W(\mathcal{P}) = W$.

We consider the space of continuous linear functionals τ on $\boldsymbol{V}(\mathcal{P}) \times W(\mathcal{P})$ that vanish on the subspace $\boldsymbol{V} \times W$. Let $\mathbb{H}(\mathbf{div}, \Omega)$ be the space

$$\mathbb{H}(\mathbf{div}, \Omega) = \left\{ \boldsymbol{A} \in L_2(\Omega)^{2 \times 2} : \mathbf{div}\, \boldsymbol{A} \in L_2(\Omega)^2 \right\}$$

equipped with the norm

$$\|A\|_{\mathbb{H}(\mathbf{div}, \Omega)} = \left\{ \|\boldsymbol{A}\|^2_{L_2(\Omega)} + \|\mathbf{div}\, \boldsymbol{A}\|^2_{L_2(\Omega)} \right\}^{1/2}.$$

The following result shows that the linear functionals that vanish on $\boldsymbol{V} \times M$ may be identified with elements of the space $\mathbb{H}(\mathbf{div}, \Omega)$. Consequently, we shall not distinguish between functionals and their representatives in $\mathbb{H}(\mathbf{div}, \Omega)$.

Theorem 9.2 *A continuous linear functional τ on the space $\boldsymbol{V}(\mathcal{P}) \times W(\mathcal{P})$ vanishes on the subspace $\boldsymbol{V} \times W$ if and only if there exists $\boldsymbol{A} \in \mathbb{H}(\mathbf{div}, \Omega)$ such that*

$$\tau(\boldsymbol{v}, q) = \sum_{K \in \mathcal{P}} \oint_{\partial K} \boldsymbol{n}_K \cdot \boldsymbol{A} \cdot \boldsymbol{v}_K \, \mathrm{d}s$$

where $\boldsymbol{n}_K$ denotes the unit outward normal vector on the boundary of K.

Proof. Any continuous functional on $H^1(K) \times H^1(K) \times L_2(K)$ may be written in the form

$$(\boldsymbol{v}, q) \to \sum_{j=1}^{2} \int_K \left\{ \sum_{i=1}^{2} A_{ij} \frac{\partial v_j}{\partial x_i} + a_j v_j \right\} \mathrm{d}\boldsymbol{x} + \int_K a_0 q \, \mathrm{d}\boldsymbol{x} \tag{9.13}$$

for suitable A_{ij}, a_j, and $a_0 \in L_2(K)$. Therefore, for any $(\boldsymbol{v}, q) \in \boldsymbol{V}(\mathcal{P}) \times W(\mathcal{P})$ we obtain

$$\tau(\boldsymbol{v}, q) = \sum_{K \in \mathcal{P}} \left[\sum_{j=1}^{2} \int_K \left\{ \sum_{i=1}^{2} A_{ij} \frac{\partial v_j}{\partial x_i} + a_j v_j \right\} \mathrm{d}\boldsymbol{x} + \int_K a_0 q \, \mathrm{d}\boldsymbol{x} \right] \tag{9.14}$$

where A_{ij}, a_j, and a_0 now denote elements of the *global* space $L_2(\Omega)$. Owing to the hypothesis on τ, it follows that for any $(\boldsymbol{v}, q) \in \boldsymbol{V} \times W$ we have

$$0 = \sum_{j=1}^{2} \int_{\Omega} \left\{ \sum_{i=1}^{2} A_{ij} \frac{\partial v_j}{\partial x_i} + a_j v_j \right\} \mathrm{d}\boldsymbol{x} + \int_{\Omega} a_0 q \, \mathrm{d}\boldsymbol{x}. \tag{9.15}$$

Hence, in the sense of distributions we obtain

$$-\sum_{i=1}^{2} \frac{\partial A_{ij}}{\partial x_i} + a_j = 0, \qquad j = 1, 2 \tag{9.16}$$

and $a_0 = 0$. Rewriting (9.16) reveals $\boldsymbol{A} \in \mathbb{H}(\mathbf{div}, \Omega)$:

$$\mathbf{div}\boldsymbol{A} = -\begin{bmatrix} a_1 \\ a_2 \end{bmatrix} \in L_2(\Omega)^2. \tag{9.17}$$

Therefore using Green's identity and (9.14), we obtain

$$\tau(\boldsymbol{v}, q) = \sum_{K \in \mathcal{P}} \sum_{i,j=1}^{2} \int_K \left\{ A_{ij} \frac{\partial v_j}{\partial x_i} + \frac{\partial A_{ij}}{\partial x_i} v_j \right\} \mathrm{d}\boldsymbol{x} = \sum_{K \in \mathcal{P}} \oint_{\partial K} \boldsymbol{n}_K \cdot \boldsymbol{A} \cdot \boldsymbol{v}_K \, \mathrm{d}s. \tag{9.18}$$

The converse is shown using similar arguments. ■

In view of the inclusion $\boldsymbol{X} \times M \subset \boldsymbol{V} \times W$, the discretization error $(\boldsymbol{e}, E)$ defined by

$$\boldsymbol{e} = \boldsymbol{u} - \boldsymbol{u}_X; \quad E = p - p_M$$

belongs to the space $\boldsymbol{V} \times W$. Define the pair $(\boldsymbol{\phi}, \psi) \in \boldsymbol{V} \times W$ such that

$$a(\boldsymbol{\phi}, \boldsymbol{v}) + c(\psi, q) = B(\boldsymbol{e}, \boldsymbol{v}) + b(\boldsymbol{v}, E) + b(\boldsymbol{e}, q) \tag{9.19}$$

for all $(\boldsymbol{v}, q) \in \boldsymbol{V} \times W$. Theorem 9.1 reveals that the norm of the discretization error is given by

$$|||(\boldsymbol{e}, E)|||^2 = ||\boldsymbol{\phi}||_a^2 + ||\psi||_c^2 .$$

The problem is therefore to estimate $||\boldsymbol{\phi}||_a$ and $||\psi||_c$ numerically. We reduce the single global problem (9.19) into a sequence of independent problems posed locally over each element.

The interelement fluxes play a vital role in defining the estimator. As in Chapter 6, we introduce a set of functions $\{\boldsymbol{g}_K : K \in \mathcal{P}\}$ that notionally approximate the normal fluxes over the element boundaries,

$$\boldsymbol{g}_K \approx \boldsymbol{n}_K \cdot \boldsymbol{\rho}(\boldsymbol{u}, p) \text{ on } \partial K,$$

where $\boldsymbol{\rho}(\boldsymbol{v}, q)$ is the stress tensor defined by

$$\rho_{ij}(\boldsymbol{v}, q) = \nu \frac{\partial v_i}{\partial x_j} - q\delta_{ij}$$

and δ_{ij} is the Kronecker symbol. The functions $\boldsymbol{g}_K \in L_2(\partial K) \times L_2(\partial K)$ are required to satisfy the consistency condition (6.7),

$$\boldsymbol{g}_K + \boldsymbol{g}_{K'} = \mathbf{0} \quad \text{on } \gamma = \partial K \cap \partial K'.$$

These conditions imply that the linear functional defined by the rule

$$\boldsymbol{V}(\mathcal{P}) \times W(\mathcal{P}) \ni (\boldsymbol{w}, q) \mapsto \sum_{K \in \mathcal{P}} \int_{\partial K} \boldsymbol{g}_K \cdot \boldsymbol{w}_K \, \mathrm{d}s$$

is continuous and, moreover, vanishes on the subspace $\boldsymbol{V} \times W$. Consequently (recalling that we identify the linear functionals of Theorem 9.2 with their representatives in $\mathbb{H}(\mathbf{div}, \Omega)$), it follows that there exists $\mu_\star \in \mathbb{H}(\mathbf{div}, \Omega)$ such that

$$\mu_\star(\boldsymbol{w}, q) = \sum_{K \in \mathcal{P}} \int_{\partial K} \boldsymbol{g}_K \cdot \boldsymbol{w}_K \, \mathrm{d}s, \quad (\boldsymbol{w}, q) \in \boldsymbol{V}(\mathcal{P}) \times W(\mathcal{P}).$$

The next step is to decompose the global problem (9.19) into local problems posed over the elements. Firstly, the unknowns $(\boldsymbol{u}, p)$ in (9.19) are replaced by appealing to (9.1):

$$\begin{aligned} & a(\boldsymbol{\phi}, \boldsymbol{w}) + c(\psi, q) \\ = \;& B(\boldsymbol{e}, \boldsymbol{w}) + b(\boldsymbol{w}, E) + b(\boldsymbol{e}, q) \\ = \;& \sum_{K \in \mathcal{P}} \{F_K(\boldsymbol{w}) - B_K(\boldsymbol{u}_X, \boldsymbol{w}) - b_K(\boldsymbol{w}, p_M) - b_K(\boldsymbol{u}_X, q)\}. \end{aligned} \tag{9.20}$$

The global space $\boldsymbol{V} \times W$ is decomposed into functions that are smooth on each of the elements but not necessarily continuous at interelement boundaries. The functional given by (9.20) is then extended to the broken space $\boldsymbol{V}(\mathcal{P}) \times W(\mathcal{P})$ as follows. For any $(\boldsymbol{w}, q) \in \boldsymbol{V}(\mathcal{P}) \times W(\mathcal{P})$, define the linear functional $\mathcal{R} : \boldsymbol{V}(\mathcal{P}) \times W(\mathcal{P}) \to \mathbb{R}$ by

$$\mathcal{R}(\boldsymbol{w}, q) = \sum_{K \in \mathcal{P}} \mathcal{R}_K(\boldsymbol{w}, q) - \mu_\star(\boldsymbol{w}, q)$$

for all $(\boldsymbol{w}, q) \in \boldsymbol{V}(\mathcal{P}) \times W(\mathcal{P})$, where

$$\begin{aligned} \mathcal{R}_K(\boldsymbol{w}, q) \;=\;& F_K(\boldsymbol{w}) - B_K(\boldsymbol{u}_X, \boldsymbol{w}) - b_K(\boldsymbol{w}, p_M) \\ -\;& b_K(\boldsymbol{u}_X, q) + \oint_{\partial K} \boldsymbol{g}_K \cdot \boldsymbol{w} \, \mathrm{d}s. \end{aligned}$$

In particular, whenever $(\boldsymbol{w}, q) \in \boldsymbol{V} \times W$, the functional assumes the form

$$\mathcal{R}(\boldsymbol{w}, q) = a(\boldsymbol{\phi}, \boldsymbol{w}) + c(\psi, q). \tag{9.21}$$

Introduce the Lagrangian functional $\mathcal{L} : \boldsymbol{V}(\mathcal{P}) \times W(\mathcal{P}) \times \mathbb{H}(\mathbf{div}, \Omega) \to \mathbb{R}$ given by

$$\mathcal{L}(\boldsymbol{w}, q, \mu) = \frac{1}{2}\{a(\boldsymbol{w}, \boldsymbol{w}) + c(q, q)\} - \mathcal{R}(\boldsymbol{w}, q) - \mu(\boldsymbol{w}, q)$$

so that

$$\sup_{\mu \in \mathbb{H}(\mathbf{div}, \Omega)} \mathcal{L}(\boldsymbol{w}, q, \mu) = \begin{cases} \frac{1}{2}\{a(\boldsymbol{w},\boldsymbol{w}) + c(q,q)\} - \mathcal{R}(\boldsymbol{w}, q) & \text{if } (\boldsymbol{w}, q) \in \boldsymbol{V} \times W \\ +\infty & \text{otherwise.} \end{cases} \tag{9.22}$$

Now, with the aid of (9.21),

$$\begin{aligned} &\frac{1}{2}\{a(\boldsymbol{w},\boldsymbol{w}) + c(q,q)\} - \mathcal{R}(\boldsymbol{w}, q) \\ &\quad = \frac{1}{2}\{a(\boldsymbol{w}-\boldsymbol{\phi}, \boldsymbol{w}-\boldsymbol{\phi}) + c(q-\psi, q-\psi)\} - \frac{1}{2}\left|\!\left|\!\left|(\boldsymbol{e}, E)\right|\!\right|\!\right|^2 \end{aligned}$$

and it follows that

$$\frac{1}{2}\{a(\boldsymbol{w},\boldsymbol{w}) + c(q,q)\} - \mathcal{R}(\boldsymbol{w}, q) = -\frac{1}{2}\left|\!\left|\!\left|(\boldsymbol{e}, E)\right|\!\right|\!\right|^2$$

for all $(\boldsymbol{w}, q) \in \boldsymbol{V} \times W$. Therefore,

$$\begin{aligned} -\frac{1}{2}\left|\!\left|\!\left|(\boldsymbol{e}, E)\right|\!\right|\!\right|^2 &= \inf_{(\boldsymbol{w},q)\in \boldsymbol{V}(\mathcal{P})\times W(\mathcal{P})} \sup_{\mu \in \mathbb{H}(\mathbf{div},\Omega)} \mathcal{L}(\boldsymbol{w}, q, \mu) \\ &= \sup_{\mu \in \mathbb{H}(\mathbf{div},\Omega)} \inf_{(\boldsymbol{w},q)\in \boldsymbol{V}(\mathcal{P})\times W(\mathcal{P})} \mathcal{L}(\boldsymbol{w}, q, \mu) \end{aligned}$$

where the change in order of the inf–sup is justified since a saddle point is obtained when the multiplier μ is the true interelement flux. The Lagrange multiplier may be selected to be $\mu_\star$, thereby yielding the bound

$$-\frac{1}{2}\left|\!\left|\!\left|(\boldsymbol{e}, E)\right|\!\right|\!\right|^2 \geq \inf_{(\boldsymbol{w},q)\in \boldsymbol{V}(\mathcal{P})\times W(\mathcal{P})} \mathcal{L}(\boldsymbol{w}, q, \mu_\star).$$

This global minimisation problem separates into independent problems for the velocity,

$$\inf_{\boldsymbol{w}\in \boldsymbol{V}(\mathcal{P})} \frac{1}{2}a(\boldsymbol{w},\boldsymbol{w}) - \mathcal{R}(\boldsymbol{w}, 0), \tag{9.23}$$

and pressure,

$$\inf_{q\in W(\mathcal{P})} \frac{1}{2}c(q,q) - \mathcal{R}(\mathbf{0}, q). \tag{9.24}$$

The fact that the broken space $\boldsymbol{V}(\mathcal{P})$ does not impose any interelement continuity requirements means that the problem (9.23) may be separated into the sum of independent problems posed over each of the element $K \in \mathcal{P}$:

$$\inf_{\boldsymbol{w}_K \in \boldsymbol{V}_K} J_K(\boldsymbol{w})$$

where $J_K : \boldsymbol{V}_K \to \mathbb{R}$ is the quadratic functional

$$\begin{aligned} J_K(\boldsymbol{w}) &= \frac{1}{2}a(\boldsymbol{w},\boldsymbol{w}) - F_K(\boldsymbol{w}) + B_K(\boldsymbol{u}_X,\boldsymbol{w}) \\ &\quad + b_K(\boldsymbol{w}, p_M) - \oint_{\partial K} \boldsymbol{g}_K \cdot \boldsymbol{w}\,\mathrm{d}s. \end{aligned}$$

The problem for the pressure may be localized in a similar fashion,

$$\sum_{K\in\mathcal{P}} \inf_{q_K\in W_K} \left\{ \frac{1}{2}\int_K q_K^2\,\mathrm{d}\boldsymbol{x} - \int_K q_K \,\mathbf{div}\,\boldsymbol{u}_X\,\mathrm{d}\boldsymbol{x} \right\},$$

apart from a single nonlocal constraint on the pressures

$$\sum_{K\in\mathcal{P}} \int_K q_K\,\mathrm{d}\boldsymbol{x} = 0. \tag{9.25}$$

The optimal choice of q_K may be found explicitly using elementary calculus,

$$q_K = \mathbf{div}\,\boldsymbol{u}_X - \lambda$$

where λ is the average value of $\mathbf{div}\,\boldsymbol{u}_X$ over the whole domain Ω,

$$\lambda \propto \sum_{K\in\mathcal{P}} \int_K \mathbf{div}\,\boldsymbol{u}_X\,\mathrm{d}\boldsymbol{x} = \oint_{\partial\Omega} \boldsymbol{n}\cdot\boldsymbol{u}_X\,\mathrm{d}s = 0.$$

As a consequence, the optimal choice is simply

$$q_K = \mathbf{div}\,\boldsymbol{u}_X$$

and the value of the objective reduces to

$$\sum_{K\in\mathcal{P}} -\frac{1}{2}\,\|\mathbf{div}\,\boldsymbol{u}_X\|^2_{L_2(K)}.$$

Summarizing these developments leads to the conclusion:

Theorem 9.3 *Let $J_K : \boldsymbol{V}_K \to \mathbb{R}$ be the quadratic functional*

$$\begin{aligned} J_K(\boldsymbol{w}) &= \frac{1}{2}a(\boldsymbol{w},\boldsymbol{w}) - F_K(\boldsymbol{w}) + B_K(\boldsymbol{u}_X,\boldsymbol{w}) \\ &\quad + b_K(\boldsymbol{w}, p_M) - \oint_{\partial K} \boldsymbol{g}_K \cdot \boldsymbol{w}\,\mathrm{d}s. \end{aligned} \tag{9.26}$$

Then,

$$|\!|\!|(e,E)|\!|\!|^2 \le \sum_{K\in\mathcal{P}} \left\{ -2 \inf_{\boldsymbol{w}_K\in\boldsymbol{V}_K} J_K(\boldsymbol{w}_K) + \|\mathbf{div}\,\boldsymbol{u}_X\|^2_{L_2(K)} \right\}. \tag{9.27}$$

The analysis has led to problems on each subdomain of the form

$$\inf_{\boldsymbol{w}_K \in \boldsymbol{V}_K} J_K(\boldsymbol{w}_K). \tag{9.28}$$

Suppose for a moment that a minimum exists, then the minimising element is characterized by finding $\boldsymbol{\phi}_K \in \boldsymbol{V}_K$ such that

$$a(\boldsymbol{\phi}_K, \boldsymbol{v}) = F_K(\boldsymbol{v}) - B_K(\boldsymbol{u}_X, \boldsymbol{v}) - b_K(\boldsymbol{v}, p_M) + \oint_{\partial K} \boldsymbol{g}_K \cdot \boldsymbol{v}\, \mathrm{d}s \quad \forall \boldsymbol{v} \in \boldsymbol{V}_K. \tag{9.29}$$

This problem decouples into a pair of Poisson type problems with Neumann data. Thus, we have shown that one can obtain a local *a posteriori* estimator for the Oseen problem by solving auxiliary *Poisson* type problems for the residual in the momentum equation. The contribution from the incompressibility constraint may be calculated explicitly. This has a considerable impact in the computation of the error estimator since one is not obliged to approximate a local Stokes-type problem—merely a Poisson problem, where issues of stability need not arise.

Necessary and sufficient conditions for the existence of a minimum are that the data satisfy the following compatibility or *equilibration* condition:

$$0 = F_K(\boldsymbol{r}) - B_K(\boldsymbol{u}_X, \boldsymbol{r}) - b_K(\boldsymbol{r}, p_M) + \oint_{\partial K} \boldsymbol{g}_K \cdot \boldsymbol{r}\, \mathrm{d}s \tag{9.30}$$

for

$$\boldsymbol{r} \in \left\{ \begin{pmatrix} 1 \\ 0 \end{pmatrix}, \begin{pmatrix} 0 \\ 1 \end{pmatrix} \right\}.$$

These conditions correspond to the equilibration condition (6.28) for each component of the momentum equation. The equilibration theory described in Chapter 6 may be applied to construct sets of boundary fluxes that respect these conditions. Specifically, the flux $\boldsymbol{g}_K$ is sought in the form

$$\boldsymbol{g}_K = \begin{pmatrix} g_K^{(1)} \\ g_K^{(2)} \end{pmatrix}$$

where each component is required to satisfy the analogue of the conditions (6.32) and (6.33). Precisely the same procedures of Chapter 6 may be employed to determine the moments of the flux needed to reconstruct the function $\boldsymbol{g}_K$ on each of the edges. It is worth noting that there is only an equilibration requirement for the momentum equations and not for the incompressibility constraint.

9.2.2 Summary

The numerical procedure consists of first calculating equilibrated boundary data for the local problems (9.29). These problems are then solved numerically, giving an approximate solution ϕ_K. The process yields an *a posteriori* error estimate η_K on the subdomain K

$$\eta_K = \left\{ \|\phi_K\|_{a,K}^2 + \|\mathbf{div}\, \boldsymbol{u}_X\|_{L_2(K)}^2 \right\}^{1/2}.$$

A global error estimate may be obtained by summing the local contributions. Theorem 9.3 guarantees that the estimate bounds the true error $|||(\boldsymbol{e}, E)|||$ from above. Two-sided bounds on the error could be obtained by arguing as in Chapter 6.

An important point of the analysis is that one does not have to solve a local Stokes problem: It is sufficient to solve a pair of independent local Poisson problems. This means that one is solving a system of two equations (since the residual corresponding to the incompressibility condition can be treated directly) rather than the system of three coupled equations needed for other techniques. Importantly, when one comes to construct the basis functions used in approximating the local problems, there is no issue of stability (inf–sup) conditions. These conditions can be quite problematic if one is trying to solve a local Stokes problem using an appropriate space, requiring a careful stability analysis [33]. This issue does not arise with the approach presented here. These features make the computation of the estimators less expensive and more easily applicable to general finite element schemes for Stokes-type problems.

Although the analysis suggests that the boundary data for the local residual-type problems should be chosen to satisfy an equilibration condition, the above comments are equally valid whether one is equilibrating the boundary fluxes or not. Of course, one loses the upper-bound property if the equilibration condition is not satisfied, but this may not be of overriding importance in some applications.

One can question the usefulness of an upper bound in the unorthodox energy norm $|||\cdot|||$, albeit equivalent with the H^1-type norms. The analysis could be used to obtain an estimator in H^1 norm (up to an unknown constant), thanks to norm equivalence of Theorem 9.1. The energy of the actual solution can be estimated in the same $|||\cdot|||$, being computed at the same time as the error estimator by modifying the right-hand sides used in the error estimation process (after omitting the terms $B_K(\boldsymbol{u}_X, \boldsymbol{v})$ and $b(\boldsymbol{v}, p_M)$ in equation (9.29)). The process yields a sufficiently good estimate for practical purposes and may be used to scale the error estimator giving an estimate of relative error.

To summarize, it has been shown that the equilibration principle carries over from the scalar case along with the basic steps in the analysis. In addition, the procedure for the treatment of side conditions, such as the incompressibility constraints, poses no additional difficulties.

9.3 INCOMPRESSIBLE NAVIER–STOKES EQUATIONS

The analysis for the Stokes and Oseen problem will be extended to the incompressible Navier–Stokes equations with small data. Let $D : \boldsymbol{V} \times \boldsymbol{V} \times \boldsymbol{V} \to \mathbb{R}$ be the trilinear form

$$D(\boldsymbol{u}, \boldsymbol{v}, \boldsymbol{w}) = \int_{\Omega} \boldsymbol{u} \cdot \boldsymbol{\nabla}\, \boldsymbol{v} \cdot \boldsymbol{w} \, \mathrm{d}\boldsymbol{x}.$$

The form D is continuous in the sense that there exists a constant C_D such that

$$D(\boldsymbol{u}, \boldsymbol{v}, \boldsymbol{w}) \leq C_D \, |\boldsymbol{u}|_{H^1(\Omega)} \, |\boldsymbol{v}|_{H^1(\Omega)} \, |\boldsymbol{w}|_{H^1(\Omega)} \,. \tag{9.31}$$

We shall assume that C_D is the best possible constant such that (9.31) holds. The incompressible Navier–Stokes problem reads as follows:

Find $(\boldsymbol{u}, p) \in \boldsymbol{V} \times W$ such that for all $(\boldsymbol{v}, q) \in \boldsymbol{V} \times W$ there holds

$$a(\boldsymbol{u}, \boldsymbol{v}) + b(\boldsymbol{v}, p) + D(\boldsymbol{u}, \boldsymbol{u}, \boldsymbol{v}) + b(\boldsymbol{u}, q) = F(\boldsymbol{v}) \tag{9.32}$$

where $F : \boldsymbol{V} \to \mathbb{R}$ is the linear functional

$$F(\boldsymbol{v}) = \int_{\Omega} \boldsymbol{f} \cdot \boldsymbol{v} \;\; \mathrm{d}\boldsymbol{x}. \tag{9.33}$$

The problem (9.33) is known [68] to possess a unique solution whenever the data is sufficiently small. In particular, if

$$|F(\boldsymbol{v})| \leq \theta \frac{\nu^2}{C_D} \, |\boldsymbol{v}|_{H^1(\Omega)} \qquad \forall \boldsymbol{v} \in \boldsymbol{V}$$

for some fixed number $\theta \in [0, 1)$, then there is a unique solution $\boldsymbol{u} \in \boldsymbol{V}$ satisfying

$$|\boldsymbol{u}|_{H^1(\Omega)} \leq \theta \frac{\nu}{C_D}. \tag{9.34}$$

The finite element approximation to (9.32) is then:

Find $(\boldsymbol{u}_X, p_M) \in \boldsymbol{X} \times M$ such that for all $(\boldsymbol{v}_X, q_M) \in \boldsymbol{X} \times M$

$$a(\boldsymbol{u}_X, \boldsymbol{v}_X) + b(\boldsymbol{v}_X, p_M) + D(\boldsymbol{u}_X, \boldsymbol{u}_X, \boldsymbol{v}_X) + b(\boldsymbol{u}_X, q_M) = F(\boldsymbol{v}_X).$$

The finite element subspaces $\boldsymbol{X}$ and M are constructed as described previously. It will be assumed that the finite element approximation $\boldsymbol{u}_X$ converges to the velocity $\boldsymbol{u}$ as the partition is refined.

Let $(\boldsymbol{e}, E)$ be the error in the finite element approximation and define a pair $(\boldsymbol{\phi}, \psi) \in \boldsymbol{V} \times W$ to be the Ritz projection of the *modified residuals*

$$a(\boldsymbol{\phi}, \boldsymbol{v}) + c(\psi, q) = a(\boldsymbol{e}, \boldsymbol{v}) + b(\boldsymbol{v}, E) + b(\boldsymbol{e}, q) + \delta(\boldsymbol{u}, \boldsymbol{u}_X, \boldsymbol{v}) \tag{9.35}$$

for all $(\boldsymbol{v}, q) \in \boldsymbol{V} \times W$, where

$$\delta(\boldsymbol{u}, \boldsymbol{u}_X, \boldsymbol{v}) = D(\boldsymbol{u}, \boldsymbol{u}, \boldsymbol{v}) - D(\boldsymbol{u}_X, \boldsymbol{u}_X, \boldsymbol{v}).$$

The data on the right-hand side of (9.35) defines a continuous linear functional and so the pair (ϕ, ψ) exists and is unique so that we may define

$$|||(e, E)||| = \left\{ \|\phi\|_a^2 + \|q\|_c^2 \right\}^{1/2}.$$

The idea echoes the basic step (9.7) used before. However, there is a significant difference: Before, the pair (e, E) was arbitrary but now it is essential that (e, E) be the error in the finite element approximation. Furthermore, owing to the presence of the nonlinear term $\delta(\cdot,\cdot,\cdot)$, one cannot directly apply Theorem 9.1. However, suppose for a moment that

$$k_1 \, |||(e, E)|||^2 \le \|e\|_a^2 + \|E\|_c^2 \le k_2 \, |||(e, E)|||^2 \tag{9.36}$$

for nonnegative quantities k_1 and k_2. Let $\phi_K \in V_K$ be such that

$$a(\phi_K, v) = F_K(v) - a_K(u_X, v) - b_K(v, p_M) - D_K(u_X, u_X, v) + \oint_{\partial K} g_K \cdot v \, ds$$

for all $v \in V_K$ where the boundary data are chosen to respect the equilibration condition,

$$0 = F_K(r) - B_K(u_X, r) - b_K(r, p_M) - D_K(u_X, u_X, r) + \oint_{\partial K} g_K \cdot r \, ds$$

where r is any constant flow field. The local *a posteriori* error estimate on element K is taken to be

$$\eta_K = \left\{ \|\phi_K\|_{a,K}^2 + \|\mathbf{div}\, u_X\|_{L_2(K)}^2 \right\}^{1/2}.$$

An argument identical to the linear case reveals that

$$|||(e, E)|||^2 \le \sum_{K \in \mathcal{P}} \eta_K^2 = \eta^2.$$

Consequently, thanks to the equivalence (9.36), η provides an error estimator for the incompressible Navier–Stokes problem with small data. It remains to prove the equivalence (9.36).

Proof. (of (9.36)): First, we obtain bounds for the form $\delta(\cdot,\cdot,\cdot)$. Suppose $v \in V$, then

$$\begin{aligned}
\delta(u, u_X, v) &= D(u, u, v) - D(u_X, u_X, v) \\
&= D(u, e, v) + D(e, u_X, v) \\
&\le C_D \, |e|_{H^1(\Omega)} \, |v|_{H^1(\Omega)} \left\{ |u|_{H^1(\Omega)} + |u_X|_{H^1(\Omega)} \right\} \\
&\le C_D \, |e|_{H^1(\Omega)} \, |v|_{H^1(\Omega)} \left\{ 2\,|u|_{H^1(\Omega)} + |e|_{H^1(\Omega)} \right\} \\
&\le C_D \, |e|_{H^1(\Omega)} \, |v|_{H^1(\Omega)} \left\{ 2\theta \frac{\nu}{C_D} + |e|_{H^1(\Omega)} \right\} \\
&= \left\{ 2\theta + \frac{C_D}{\nu} \, |e|_{H^1(\Omega)} \right\} \|e\|_a \, \|v\|_a
\end{aligned} \tag{9.37}$$

where (9.31) and (9.34) have been used. If $\boldsymbol{v} = \boldsymbol{e}$, then a sharper bound may be obtained by first noting that [68, equation (6), page 285]

$$D(\boldsymbol{u}, \boldsymbol{e}, \boldsymbol{e}) = 0$$

and then following similar steps before giving

$$\delta(\boldsymbol{u}, \boldsymbol{u}_X, \boldsymbol{e}) \le \left\{ \theta + \frac{C_D}{\nu} |\boldsymbol{e}|_{H^1(\Omega)} \right\} \|\boldsymbol{e}\|_a^2 . \tag{9.38}$$

Left Hand Inequality. Using (9.4), (9.35), and (9.37) gives

$$\|\boldsymbol{\phi}\|_a \le \nu^{-1/2} \|E\|_c + \left\{ 3 + \frac{C_D}{\nu} |\boldsymbol{e}|_{H^1(\Omega)} \right\} \|\boldsymbol{e}\|_a .$$

Using (9.4) and (9.35) gives

$$\|\psi\|_c \le \nu^{-1/2} \|\boldsymbol{e}\|_a$$

and hence

$$k_1 \left\{ 1 + \mathcal{O}[|\boldsymbol{e}|_{H^1(\Omega)}] \right\} |||(\boldsymbol{e}, E)|||^2 \le \|\boldsymbol{e}\|_a^2 + \|E\|_c^2$$

where k_1 depends on C_D and ν.

Right Hand Inequality. Using (9.5), (9.35) and (9.37) gives

$$\begin{aligned} \alpha_b \|E\|_c &\le \sup_{\boldsymbol{v} \in \boldsymbol{V}} \frac{|b(\boldsymbol{v}, q)|}{\|\boldsymbol{v}\|_a} \\ &\le \sup_{\boldsymbol{v} \in \boldsymbol{V}} \frac{1}{\|\boldsymbol{v}\|_a} |a(\boldsymbol{\phi}, \boldsymbol{v}) - a(\boldsymbol{e}, \boldsymbol{v}) - \delta(\boldsymbol{u}, \boldsymbol{u}_X, \boldsymbol{v})| \\ &\le \|\boldsymbol{\phi}\|_a + \|\boldsymbol{e}\|_a \left\{ 3 + \frac{C_D}{\nu} |\boldsymbol{e}|_{H^1(\Omega)} \right\} . \end{aligned} \tag{9.39}$$

Equally well, using (9.4), (9.35), and (9.38) gives

$$\begin{aligned} \|\boldsymbol{e}\|_a^2 &= a(\boldsymbol{\phi}, \boldsymbol{e}) - b(\boldsymbol{e}, E) - \delta(\boldsymbol{u}, \boldsymbol{u}_X, \boldsymbol{e}) \\ &= a(\boldsymbol{\phi}, \boldsymbol{e}) - c(\psi, E) - \delta(\boldsymbol{u}, \boldsymbol{u}_X, \boldsymbol{e}) \\ &\le \|\boldsymbol{\phi}\|_a \|\boldsymbol{e}\|_a + \|\psi\|_c \|E\|_c + \left\{ \theta + \frac{C_D}{\nu} |\boldsymbol{e}|_{H^1(\Omega)} \right\} \|\boldsymbol{e}\|_a^2 . \end{aligned} \tag{9.40}$$

Since it is assumed that $|\boldsymbol{e}|_{H^1(\Omega)} \to 0$, we may choose $\epsilon > 0$ sufficiently small that

$$\theta + \frac{\epsilon^2}{2} + \frac{C_D}{\nu} |\boldsymbol{e}|_{H^1(\Omega)} = \theta' < 1.$$

Hence, from (9.40) and (9.39) we obtain

$$\|\boldsymbol{e}\|_a^2 \le \frac{1}{\alpha_b} \|\boldsymbol{\phi}\|_a \|\psi\|_c + \|\boldsymbol{e}\|_a \left\{ \|\boldsymbol{\phi}\|_a + \frac{1}{\alpha_b} \left(3 + \frac{C_D}{\nu} |\boldsymbol{e}|_{H^1(\Omega)} \right) \|\psi\|_c \right\}$$

$$
\begin{aligned}
&+ \|e\|_a^2 \left\{ \theta + \frac{C_D}{\nu} |e|_{H^1(\Omega)} \right\} \\
\leq \; & \frac{1}{\alpha_b} \|\phi\|_a \|\psi\|_c + \theta' \|e\|_a^2 \\
&+ \frac{1}{2\epsilon^2} \left\{ \|\phi\|_a + \frac{1}{\alpha_b} \left(3 + \frac{C_D}{\nu} |e|_{H^1(\Omega)} \right) \|\psi\|_c \right\}^2 . \qquad (9.41)
\end{aligned}
$$

Hence, for $|e|_{H^1(\Omega)}$ sufficiently small we obtain

$$
\|e\|_a^2 \leq C(\alpha_b, \nu, C_D, |e|_{H^1(\Omega)}, \theta) \left\{ \|\phi\|_a^2 + \|\psi\|_c^2 \right\}
$$

and using (9.38) gives

$$
\|E\|_c^2 \leq C(\alpha_b, \nu, C_D, |e|_{H^1(\Omega)}, \theta) \left\{ \|\phi\|_a^2 + \|\psi\|_c^2 \right\},
$$

and the result follows. ∎

9.4 EXTENSIONS TO NONLINEAR PROBLEMS

9.4.1 A Class of Nonlinear Problems

Our goal is to provide an informal indication of how the methods of *a posteriori* error estimation extend to certain nonlinear boundary value problems. The basic idea may be described by appealing to the following class of nonlinear boundary value problem:

$$
-\boldsymbol{\nabla} \cdot (a(\boldsymbol{x}, u) \boldsymbol{\nabla} u) = f(\boldsymbol{x}, u) \text{ in } \Omega
$$

subject to the boundary condition $u = 0$ on $\partial\Omega$. The variational form of this problem consists of finding $u \in V$ such that

$$
B(u; u, w) = L(u; w) \quad \forall w \in W \qquad (9.42)
$$

where V and W are appropriate reflexive Banach spaces. Typically, these will be related to non-Hilbertian Sobolev spaces so that $V = W_0^{1,p}(\Omega)$ and $W = W_0^{1,q}(\Omega)$ for suitably chosen values of p, $q \in (1, \infty)$, depending on the specific form of the functions a and f. The forms are given by

$$
B(u; v, w) = \int_\Omega a(\boldsymbol{x}, u) \boldsymbol{\nabla} v \cdot \boldsymbol{\nabla} w \, \mathrm{d}\boldsymbol{x}
$$

and

$$
L(u; w) = \int_\Omega f(\boldsymbol{x}, u) w \, \mathrm{d}\boldsymbol{x}.
$$

The coefficients are assumed to satisfy $a \in C^1(\overline{\Omega} \times \mathbb{R})$, $f \in C(\overline{\Omega} \times \mathbb{R})$ and it is assumed that there exists a constant α such that

$$a(\boldsymbol{x}, y) \geq \alpha > 0 \quad \text{for } \boldsymbol{x} \in \Omega, y \in \mathbb{R}.$$

Let $X \subset V \cap W$ be the usual finite element space constructed on a regular partition $\mathcal{P}$, and define the finite element approximation $u_X \in X$ by

$$B(u_X; u_X, w_X) = L(u_X; w_X) \quad \forall w_X \in X. \tag{9.43}$$

It is assumed that both the original problem (9.42) and its finite element approximation (9.43) have unique solutions.

The residual operator $\mathcal{R} : V \to W^*$ is defined by

$$(\mathcal{R}(v), w) = L(v; w) - B(v; v, w), \quad v \in V, w \in W.$$

Observe, in particular, that

$$(\mathcal{R}(u), w) = 0 \quad \forall w \in W \tag{9.44}$$

and

$$(\mathcal{R}(u_X), w_X) = 0 \quad \forall w_X \in X. \tag{9.45}$$

The derivative of the residual operator at $v \in V$ is defined for each $\phi \in V$ and $w \in W$ by

$$(D\mathcal{R}(v)\phi, w) = \lim_{\epsilon \to 0} \frac{1}{\epsilon}(\mathcal{R}(v + \epsilon\phi) - \mathcal{R}(v), w),$$

and as a consequence we obtain

$$\begin{aligned} (D\mathcal{R}(v)\phi, w) &= \int_\Omega \frac{\partial f}{\partial y}(\boldsymbol{x}, v)\phi w \, d\boldsymbol{x} \\ &- \int_\Omega a(\boldsymbol{x}, v)\boldsymbol{\nabla}\phi \cdot \boldsymbol{\nabla} w \, d\boldsymbol{x} - \int_\Omega \frac{\partial a}{\partial y}(\boldsymbol{x}, v)\phi \boldsymbol{\nabla} v \cdot \boldsymbol{\nabla} w \, d\boldsymbol{x}. \end{aligned}$$

Under suitable conditions on the data a and f, the derivative $D\mathcal{R}$ belongs to the space $\mathcal{L}(V, W^*)$ of bounded linear mappings from V to W^*. If, in addition, there is a positive constant C_0 such that for all $F \in W^*$ there exists a unique $\phi \in V$ such that

$$(D\mathcal{R}(u)\phi, w) = (F, w) \quad \forall w \in W$$

with

$$\|\phi\|_V \leq C_0 \|F\|_{W^*},$$

then $u \in V$ is said to be a *regular* solution of the problem (9.42). Finally, if there exist positive constants C_1 and ε such that for all v with $\|v\|_V < \varepsilon$ we obtain

$$\|D\mathcal{R}(u) - D\mathcal{R}(u + tv)\|_{\mathcal{L}(V,W^*)} \leq C_1 t \|v\|_V, \quad t \in [0, 1],$$

then $D\mathcal{R}$ is said to be *Lipschitz in an ε-neighborhood of u*.

9.4.2 A Posteriori Error Estimation

Let $e = u - u_X \in V$ be the error in the finite element approximation. Then,

$$(\mathcal{R}(u_X), w) = (\mathcal{R}(u), w) + \int_0^1 (D\mathcal{R}(u + te)e, w)\,\mathrm{d}t$$

and thanks to (9.44) we have

$$(\mathcal{R}(u_X), w) = \int_0^1 (D\mathcal{R}(u + te)e, w)\,\mathrm{d}t$$

for $w \in W$. Hence,

$$(D\mathcal{R}(u)e, w) = (F, w) \tag{9.46}$$

where

$$(F, w) = \int_0^1 (D\mathcal{R}(u)e - D\mathcal{R}(u + te)e, w)\,\mathrm{d}t + (\mathcal{R}(u_X), w), \tag{9.47}$$

and if u is a regular solution, then

$$\|e\|_V \leq C_0 \|F\|_{W^*}.$$

The quantity appearing on the right-hand side may be estimated as follows. Firstly, note that

$$\left|\int_0^1 (D\mathcal{R}(u)e - D\mathcal{R}(u + te)e, w)\,\mathrm{d}t\right|$$
$$\leq \int_0^1 \|D\mathcal{R}(u)e - D\mathcal{R}(u + te)e\|_{W^*} \|w\|_W \,\mathrm{d}t$$

and

$$\|D\mathcal{R}(u)e - D\mathcal{R}(u + te)e\|_{W^*} \leq t \|D\mathcal{R}(u) - D\mathcal{R}(u + te)\|_{\mathcal{L}(V,W^*)} \|e\|_V.$$

If the derivative of $\mathcal{R}$ is Lipschitz in an ε-neighborhood of u, then, for $\|e\|_V < \varepsilon$,

$$\left|\int_0^1 (D\mathcal{R}(u)e - D\mathcal{R}(u + te)e, w)\,\mathrm{d}t\right| \leq \frac{1}{2} C_1 \|e\|_V^2 \|w\|_W.$$

Hence,

$$\|F\|_{W^*} \leq \|\mathcal{R}(u_X)\|_{W^*} + \frac{1}{2} C_1 \|e\|_V^2.$$

Consequently,

$$\|e\|_V \leq C_0 \|\mathcal{R}(u_X)\|_{W^*} + \frac{1}{2} C_0 C_1 \|e\|_V^2$$

and if $\|e\|_V$ is sufficiently small, say

$$\|e\|_V \leq 1/C_0 C_1,$$

then

$$\|e\|_V \leq 2C_0 \|\mathcal{R}(u_X)\|_{W^*}.$$

In summary, we have proved the following:

Lemma 9.4 *Suppose that u is a regular solution of the problem (9.42) and let e denote the error in the finite element approximation. If $D\mathcal{R}$ is Lipschitz in an ε-neighborhood of u and if the error e is sufficiently small, then*

$$\frac{1}{C} \|\mathcal{R}(u_X)\|_{W^*} \leq \|e\|_V \leq C \|\mathcal{R}(u_X)\|_{W^*}$$

for a positive constant C.

Proof. The upper bound follows at once from the foregoing arguments. Equations (9.46) and (9.47) imply that

$$(\mathcal{R}(u_X), w) = (D\mathcal{R}(u_X)e, w) - \int_0^1 (D\mathcal{R}(u)e - D\mathcal{R}(u+te)e, w)\,\mathrm{d}t, \quad w \in W.$$

Using the bounds derived earlier leads to the conclusion

$$\|\mathcal{R}(u_X)\|_{W^*} \leq \|D\mathcal{R}(u_X)\|_{\mathcal{L}(V,W^*)} \|e\|_V + \frac{1}{2} C_1 \|e\|_V^2.$$

Thanks to $D\mathcal{R}$ being Lipschitz, the left-hand bound follows at once provided that e is sufficiently small. ∎

9.4.3 Estimation of the Residual

It remains to estimate the residual $\mathcal{R}(u_X)$. Suppose that the space W is given by $W_0^{1,q}(\Omega)$, and observe that for $w \in W$ we have

$$\begin{aligned}(\mathcal{R}(u_X), w) &= L(u_X; w) - B(u_X; u_X, w) \\ &= \sum_{K \in \mathcal{P}} \left(\int_K f(\boldsymbol{x}, u_X) w \,\mathrm{d}\boldsymbol{x} - \int_K a(\boldsymbol{x}, u_X) \nabla u_X \cdot \nabla w \,\mathrm{d}\boldsymbol{x} \right).\end{aligned}$$

Now, following the approach of Chapter 6, we introduce a set of approximate fluxes $\{g_K : K \in \mathcal{P}\}$ satisfying the conditions (6.7) and (6.8). The residual may then be decomposed into local contributions,

$$(\mathcal{R}(u_X), w) = \sum_{K \in \mathcal{P}} (\mathcal{R}_K(u_X), w)_K$$

where

$$(\mathcal{R}_K(u_X), w)_K$$
$$= \int_K f(\boldsymbol{x}, u_X) w \, \mathrm{d}\boldsymbol{x} - \int_K a(\boldsymbol{x}, u_X) \boldsymbol{\nabla} u_X \cdot \boldsymbol{\nabla} w \, \mathrm{d}\boldsymbol{x} + \int_{\partial K} g_K w \, \mathrm{d}s,$$

is the residual linearised around the approximate solution u_X. Define $\phi_K \in V_K$ to be the solution of the (linear) local residual problem

$$B_K(u_X; \phi_K, w) = (\mathcal{R}_K(u_X), w)_K \quad \forall w \in W_K, \tag{9.48}$$

so that

$$(\mathcal{R}(u_X), w) = \sum_{K \in \mathcal{P}} B_K(u_X; \phi_K, w).$$

The steps leading to Theorem 6.1 may now be followed to obtain an upper bound for the residual. In particular, with the aid of Hölder's inequality we obtain

$$\begin{aligned} B_K(u_X; \phi_K, w) &= \int_K a(\boldsymbol{x}, u_X) \boldsymbol{\nabla} \phi_K \cdot \boldsymbol{\nabla} w \, \mathrm{d}\boldsymbol{x} \\ &\leq \|a(\cdot, u_X) \boldsymbol{\nabla} \phi_K\|_{L_{q'}(K)} \|\boldsymbol{\nabla} w\|_{L_q(K)} \end{aligned}$$

where

$$\frac{1}{q'} + \frac{1}{q} = 1.$$

Summing over the elements in the partition and applying Hölder's inequality once again will lead to the conclusion

$$|(\mathcal{R}(u_X), w)| \leq \left(\sum_{K \in \mathcal{P}} \|a(\cdot, u_X) \boldsymbol{\nabla} \phi_K\|_{L_{q'}(K)}^{q'} \right)^{1/q'} \|w\|_W,$$

and hence

$$\|\mathcal{R}(u_X)\|_{W^*} \leq \left(\sum_{K \in \mathcal{P}} \|a(\cdot, u_X) \boldsymbol{\nabla} \phi_K\|_{L_{q'}(K)}^{q'} \right)^{1/q'}.$$

In summary:

Theorem 9.5 *Suppose that the conditions of Lemma 9.4 hold, with the space W chosen to be $W = W_0^{1,q}(\Omega)$. Let ϕ_K be the solution of the linearised residual problem (9.48); then*

$$\|e\|_V \leq C \left(\sum_{K \in \mathcal{P}} \|a(\cdot, u_X) \boldsymbol{\nabla} \phi_K\|_{L_{q'}(K)}^{q'} \right)^{1/q'}$$

where C is the same positive constant as in Lemma 9.4.

9.5 BIBLIOGRAPHICAL REMARKS

Explicit *a posteriori* error estimates for the Stokes problem have been derived by Baranger and ElAmri [38] and Verfürth [109]. Generalizations of the classical element residual method have been developed by Bank and Welfert [33] and Verfürth [109] and involve the solution of local Stokes problems. The discussion given in the text follows that of Ainsworth and Oden [12]. The discussion of the Navier–Stokes problem is based on reference [94]. The extension to nonlinear problems is based on Verfürth [110]. For further details on the formulation of nonlinear problems and related concepts, the reader is referred to reference [118].

References

1. B. Achchab and J.F. Maitre. Estimate of the constant in two strengthened Cauchy Schwarz Buniakowsky inequalities for FEM systems of two dimensional elasticity–application to multilevel methods and a-posteriori error estimators. *Num. Lin. Alg. with Applic.*, 3(2):147–159, 1996.

2. M. Ainsworth. The performance of Bank–Weiser's error estimator for quadrilateral finite elements. *Numer. Methods PDE.*, 10:609–623, 1994.

3. M. Ainsworth. The influence and selection of subspaces for a posteriori error estimators. *Numer. Math.*, 73:399–418, 1996.

4. M. Ainsworth. Identification and a posteriori estimation of pollution errors in finite element analysis. *Comput. Methods Appl. Mech. Engrg.*, 176:3–17, 1999.

5. M. Ainsworth and I. Babuška. Reliable and robust a posteriori error estimation for singularly perturbed reaction diffusion problems. *SIAM J. Numer. Anal.*, 36(2):331–353, 1999.

6. M. Ainsworth and A.W. Craig. A posteriori error estimators in the finite element method. *Numer. Math.*, 60:429–463, 1991.

7. M. Ainsworth, L. Demkowicz, and C. Kim. The equilibrated residual method for a posteriori error estimation on one-irregular *hp*-finite element meshes in two dimensions. *Preprint*, 2000.

8. M. Ainsworth and J.T. Oden. A procedure for a posteriori error estimation for h–p finite element methods. *Comput. Methods Appl. Mech. Engrg.*, 101:73–96, 1992.

9. M. Ainsworth and J.T. Oden. A posteriori error estimators for second order elliptic systems. Part 1: Theoretical foundations and a posteriori error analysis. *Comput. Math. Appl.*, 25:101–113, 1993.

10. M. Ainsworth and J.T. Oden. A posteriori error estimators for second order elliptic systems. Part 2: An optimal order process for calculating self equilibrating fluxes. *Comput. Math. Appl.*, 26:75–87, 1993.

11. M. Ainsworth and J.T. Oden. A unified approach to a posteriori error estimation based on element residual methods. *Numer. Math.*, 65:23–50, 1993.

12. M. Ainsworth and J.T. Oden. A posteriori error estimates for Stokes' and Oseen's equations. *SIAM J. Numer. Anal.*, 34(1):228–245, February 1997.

13. M. Ainsworth and J.T. Oden. A posteriori error estimation in finite element analysis. *Comput. Methods Appl. Mech. Engrg.*, 142(1-2):1–88, 1997.

14. M. Ainsworth and B. Senior. Aspects of an adaptive *hp*-finite element method: Adaptive strategy, conforming approximation and efficient solvers. *Comput. Methods Appl. Mech. Engrg.*, 150:65–87, 1997.

15. M. Ainsworth and B. Senior. *hp*-finite element procedures on non-uniform geometric meshes: Adaptivity and constrained approximation. In M.W. Bern, J.E. Flaherty, and M. Luskin, editors, *Grid Generation and Adaptive Algorithms*, volume 113, pages 1–29. IMA, Minnesota, 1999.

16. I. Babuška and A. Miller. The post-processing approach in the finite element method. Part 1. Calculation of displacements, stresses and other higher derivatives of the displacements. *Internat. J. Numer. Methods Engrg.*, 20:1085–1109, 1984.

17. I. Babuška and A. Miller. The post-processing approach in the finite element method. Part 2. The calculation of stress intensity factors. *Internat. J. Numer. Methods Engrg.*, 20:1111–1129, 1984.

18. I. Babuška and A. Miller. The post-processing approach in the finite element method. Part 3. A posteriori error estimates and adaptive mesh selection. *Internat. J. Numer. Methods Engrg.*, 20:2311–2324, 1984.

19. I. Babuška and A. Miller. A feedback finite element method with a posteriori error estimation Part 1. *Comput. Methods Appl. Mech. Engrg.*, 61:1–40, 1987.

20. I. Babuška and W. C. Rheinboldt. Error estimates for adaptive finite element computations. *SIAM J. Numer. Anal.*, 18:736–754, 1978.

21. I. Babuška and W. C. Rheinboldt. A posteriori error analysis of finite element solutions for one dimensional problems. *SIAM J. Numer. Anal.*, 18:565–589, 1981.

22. I. Babuška and W.C. Rheinboldt. A posteriori error estimates for the finite element method. *Internat. J. Numer. Methods Engrg.*, 12:1597–1615, 1978.

23. I. Babuška and W.C. Rheinboldt. Adaptive approaches and reliability estimations in finite element analysis. *Comput. Methods Appl. Mech. Engrg.*, 17:519–540, 1979.

24. I. Babuška and W.C. Rheinboldt. Analysis of optimal finite element meshes in R^1. *Math. Comp.*, 33:435–463, 1979.

25. I. Babuška, T. Strouboulis, S.K. Gangaraj, K. Copps, and D.K. Datta. A posteriori estimation of the error in the error estimate. In Ladevèze and Oden [78], pages 155–198.

26. I. Babuška, T. Strouboulis, S.K. Gangaraj, and C.S. Upadhyay. Eta percent superconvergence in the interior of locally refined meshes of quadrilaterals-superconvergence of the gradient in finite element solutions of Laplace and Poisson equations. *Appl. Numer. Math.*, 16(1–2):3–49, 1994.

27. I. Babuška, T. Strouboulis, S.K. Gangaraj, and C.S. Upadhyay. Pollution error in the h-version of the finite element method and the local quality of the recovered derivatives. *Comput. Methods Appl. Mech. Engrg.*, 140(1-2):1–37, 1997.

28. I. Babuška, T. Strouboulis, and C.S. Upadhyay. Eta percent superconvergence of finite element approximations in the interior of general meshes of triangles. *Comput. Methods Appl. Mech. Engrg.*, 122(3-4):273–305, 1995.

29. I. Babuška, T. Strouboulis, C.S. Upadhyay, and S.K. Gangaraj. A model study of the quality of a posteriori estimators for linear elliptic problems error estimation in the interior of patchwise uniform grids of triangles. *Comput. Methods Appl. Mech. Engrg.*, 114:307–378, 1994.

30. I. Babuška, T. Strouboulis, C.S. Upadhyay, and S.K. Gangaraj. A posteriori estimation and adaptive-control of the pollution error in the h-version of the finite-element method. *Internat. J. Numer. Methods Engrg.*, 38(24):4207–4235, 1995.

31. I. Babuška, T. Strouboulis, C.S. Upadhyay, S.K. Gangaraj, and K. Copps. Validation of a posteriori error estimators by a numerical approach. *Internat. J. Numer. Methods Engrg.*, 37:1073–1123, 1994.

32. I. Babuška, O.C. Zienkiewicz, J. Gago, and E.A. Oliveira. *Accuracy Estimates and Adaptive Refinements in Finite Element Computations.* Wiley, 1986.

33. R. Bank and B.D. Welfert. A posteriori error estimates for the Stokes problem. *SIAM J. Numer. Anal.*, 28:591–623, 1991.

34. R.E. Bank. Analysis of a local a posteriori error estimat for elliptic equations. In *Accuracy Estimates and Adaptive Refinements in Finite Element Computations* [32], pages 119–128.

35. R.E. Bank. Hierarchical bases and the finite element method. *Acta Numerica*, 5:1–45, 1996.

36. R.E. Bank and R.K. Smith. A posteriori error-estimates based on hierarchical bases. *SIAM J. Numer. Anal.*, 30(4):921–935, 1993.

37. R.E. Bank and A. Weiser. Some a posteriori error estimators for elliptic partial differential equations. *Math. Comp.*, 44:283–301, 1985.

38. J. Baranger and H. El-Amri. Estimateurs a posteriori d'erreur pour le calcul adaptatif d'écoulements quasi–Newtoniens. *RAIRO Anal. Numér.*, 25:31–48, 1991.

39. J. Bergh and J. Löfström. *Interpolation Spaces: An introduction.* Springer-Verlag, 1976.

40. C. Bernardi and V. Girault. A local regularisation operator for triangular and quadrilateral finite elements. *SIAM J. Numer. Anal.*, 35(5):1893–1916, October 1998.

41. F. Bornemann and H. Yserentant. A basic norm equivalence in the theory of multilevel methods. *Numer. Math.*, 64(4):455–476, 1993.

42. F.A. Bornemann, B. Erdmann, and R. Kornhuber. A-posteriori error-estimates for elliptic problems in 2 and 3 space dimensions. *SIAM J. Numer. Anal.*, 33(3):1188–1204, 1996.

43. B. Braess and R. Verfürth. Posteriori error estimators for the Raviart Thomas element. *SIAM J. Numer. Anal.*, 33(6):2431–2444, 1996.

44. D. Braess. The contraction number of a multigrid method for solving the Poisson equation. *Numer. Math.*, 37:387–404, 1981.

45. S.C. Brenner and L.R. Scott. *The Mathematical Theory of Finite Element Methods*, volume 15 of *Texts in Applied Mathematics.* Springer-Verlag, New York, 1994.

46. J.C. Butcher. *The numerical analysis of ordinary differential equations: Runge-Kutta and general linear methods.* Wiley-Interscience. John Wiley & Sons, 1987.

47. T. Cao, D.W. Kelly, and I.H. Sloan. Local error bounds for post-processed finite element calculations. *Internat. J. Numer. Methods Engrg.*, 45(8):1085–1098, 1999.

48. C. Carstensen. A posteriori error estimate for the mixed finite element method. *Math. Comp.*, 66(218):465–476, 1997.

49. C. Carstensen and G. Dolzmann. A posteriori error estimates for mixed finite element method in elasticity. *Numer. Math.*, 81(2):187–209, 1998.

50. P.G. Ciarlet. *The Finite Element Method for Elliptic Problems.* North–Holland, 1978.

51. P. Clément. Approximation by finite element functions using local regularization. *RAIRO Anal. Numér.*, 2:77–84, 1975.

52. B. Fraeijs de Veubeke. Displacement and equilibrium models in the finite element method. In Zienkiewicz and Holister, editors, *Stress Analysis.* Wiley London, 1965.

53. L. Demkowicz, Ph. Devloo, and J.T. Oden. On an h-type mesh refinement strategy based on minimization of interpolation errors. *Comput. Methods Appl. Mech. Engrg.*, 53:67–89, 1985.

54. L. Demkowicz, J.T. Oden, W. Rachowicz, and O. Hardy. Toward a universal *hp*-adaptive finite element strategy. Part 1: Constrained approximation and data structure. *Comput. Methods Appl. Mech. Engrg.*, 77:79–112, 1989.

55. L. Demkowicz, J.T. Oden, and T. Strouboulis. Adaptive finite elements for flow problems with moving boundaries. Part 1: Variational principles and a posteriori error estimates. *Comput. Methods Appl. Mech. Engrg.*, 46:217–251, 1984.

56. L. Demkowicz, J.T. Oden, and T. Strouboulis. An adaptive p-version finite element method for transient flow problems with moving boundaries. In R.H. Gallagher, editor, *Finite Elements in Fluids VI*, pages 291–305. John Wiley, 1985.

57. P. Deuflhard, P. Leinen, and H. Yserentant. Concepts of an adaptive hierarchical finite element code. *IMPACT of Comput. in Sci. and Eng.*, 1:3–35, 1989.

58. W. Dörfler. A robust adaptive strategy for the nonlinear Poisson equation. *Computing*, 55(4):289–304, 1995.

59. W. Dörfler. A convergent adaptive algorithm for Poisson's equation. *SIAM J. Numer. Anal.*, 33(3):1106–1124, 1996.

60. T.F. Dupont and L.R. Scott. Polynomial approximation of functions in sobolev spaces. *Math. Comp.*, 1980:441–463, 34.

61. R. Duran, E.A. Dari, and C. Padra. Maximum norm error estimators for three dimensional elliptic problems. *SIAM J. Numer. Anal.*, 37(2):683–700, 2000.

62. R. Duran and R. Rodriguez. On the asymptotic exactness of Bank-Weiser's estimator. *Numer. Math.*, 62:297–304, 1992.

63. V. Eijkhout and P. Vassilevski. The role of the strengthened Cauchy–Buniakowskii–Schwarz inequality in multilevel methods. *SIAM Rev.*, 33(3):405–419, 1991.

64. K. Eriksson and C. Johnson. Error-estimates and automatic time step control for nonlinear parabolic problems. Part 1. *SIAM J. Numer. Anal.*, 24(1):12–23, 1987.

65. K. Eriksson and C. Johnson. Adaptive finite-element methods for parabolic problems. Part 1: A linear-model problem. *SIAM J. Numer. Anal.*, 28(1):43–77, 1991.

66. K. Eriksson and C. Johnson. Adaptive finite-element methods for parabolic problems. Part 2: Optimal error estimates in $L_\infty L_2$ and $L_\infty L_\infty$. *SIAM J. Numer. Anal.*, 32(3), 1995.

67. R.E. Ewing. A posteriori error estimation. *Comput. Methods Appl. Mech. Engrg.*, 82(1-3):59–72, 1990.

68. V. Girault and P.A. Raviart. *Finite Element Methods for Navier Stokes Equations*, volume 5 of *Springer Series in Computational Mathematics.* Springer-Verlag, 1986.

69. P. Grisvard. *Elliptic Problems in Non-smooth Domains*, volume 24 of *Monographs and Studies in Mathematics.* Pitman, 1985.

70. E. Hairer, S.P. Norsett, and G. Wanner. *Solving Ordinary Differential Equations, volume 1: Non-stiff problems*, volume 8 of *Springer Series in Computational Mathematics.* Springer-Verlag, 1987.

71. P. Houston, J.A. Mackenzie, E. Suli, and G. Warnecke. A posteriori error analysis for numerical approximations of Friedrichs systems. *Numer. Math.*, 82(3):433–470, 1999.

72. C. Johnson and P. Hansbo. Adaptive finite element methods in computational mechanics. *Comput. Methods Appl. Mech. Engrg.*, 101(1-3):143–181, 1992.

73. D.W. Kelly. The self-equilibration of residuals and complementary a posteriori error estimates in the finite element method. *Internat. J. Numer. Methods Engrg.*, 20:1491–1506, 1984.

74. D.W. Kelly, J.R. Gago, O.C. Zienkiewicz, and I. Babuška. A posteriori error analysis and adaptive processes in the finite element method. Part I–Error analysis. *Internat. J. Numer. Methods Engrg.*, 19:1593–1619, 1983.

75. D.W. Kelly and J. D. Isles. A procedure for a posteriori error analysis for the finite element method which contains a bounding measure. *Comput. Struct.*, 31:63–71, 1989.

76. M. Križek and P. Neitaanmaki. On superconvergence techniques. *Acta Applicandae Mathematicae*, 9:175–198, 1987.

77. P. Ladevèze and D. Leguillon. Error estimate procedure in the finite element method and applications. *SIAM J. Numer. Anal.*, 20:485–509, 1983.

78. P. Ladevèze and J.T. Oden, editors. *Advances in Adaptive Computational Methods in Mechanics*, volume 47, North-Holland, 1998. Elsevier Science Publishers B.V.

79. P. Lesaint and M. Zlamal. Superconvergence of the gradient of finite element solutions. *RAIRO Anal. Numér.*, 13:139–166, 1979.

80. N. Levine. Superconvergent recovery of the gradient from piecewise linear finite element approximations. *IMA J. Numer. Anal.*, 5(4):407–427, 1985.

81. J.L. Lions and E. Magenes. *Non-homogeneous boundary value problems and applications–I*, volume 181 of *Die Grundlehren der mathematischen Wissenschaften*. Springer-Verlag, 1972.

82. A. Louis. Acceleration of convergence for finite element solutions of the Poisson equation. *Numer. Math.*, 33:43–53, 1979.

83. J.F. Maitre and F. Musy. The contraction number of a class of two-level methods; an exact evaluation for some finite element subspaces and model problems. In *Multigrid Methods: Proceedings Cologne 1981*, volume 960 of *Lecture Notes in Mathematics*, pages 535–544. Springer-Verlag, 1982.

84. R.H. Nochetto. Removing the saturation assumption in finite element analysis. *Istit. Lombardo Acad. Sci. Lett. Rend. A*, 127:67–82, 1993.

85. R.H. Nochetto. Pointwise a posteriori error estimates for elliptic problems on highly graded meshes. *Math. Comp.*, 64(109):1–22, 1995.

86. A. K. Noor and I. Babuška. Quality assessment and control of finite element solutions. *Finite Elements and Design*, 3:1–26, 1987.

87. J.T. Oden and L. Demkowicz. Advances in adaptive improvements: A survey of adaptive finite element methods in computational mechanics. In *Accuracy Estimates and Adaptive Refinements in Finite Element Computations* [32], pages 1–43.

88. J.T. Oden, L. Demkowicz, W. Rachowicz, and T.A. Westermann. Toward a universal *hp*-adaptive finite element strategy. Part 2: A posteriori error estimation. *Comput. Methods Appl. Mech. Engrg.*, 77:113–180, 1989.

89. J.T. Oden, L. Demkowicz, T. Strouboulis, and Ph. Devloo. Adaptive methods for problems in solid and fluid mechanics. In *Accuracy Estimates and Adaptive Refinements in Finite Element Computations* [32], pages 249–280.

90. J.T. Oden and Y. Feng. Local and pollution error estimation for finite element approximations of elliptic boundary value problems. *J. Comput. Appl. Math.*, 74:245–293, 1996.

91. J.T. Oden and S. Prudhomme. A technique for a posteriori error estimation of h–p approximations of the Stokes equations. In Ladevèze and Oden [78], pages 43–63.

92. J.T. Oden and S. Prudhomme. Goal oriented error estimation and adaptivity for the finite element method. *Comput. Math. Appl.*, To appear.

93. J.T. Oden and J.N. Reddy. *An introduction to the mathematical theory of finite elements.* Wiley, 1976.

94. J.T. Oden, W. Wu, and M. Ainsworth. A posteriori error estimators for the Navier-Stokes problem. *Comput. Methods Appl. Mech. Engrg.*, 111:185–202, 1994.

95. M. Paraschivoiu, J. Peraire, and A.T. Patera. A posteriori finite element bounds for linear functional outputs of elliptic partial differential equations. *Comput. Methods Appl. Mech. Engrg.*, 150(1-4):289–312, 1997.

96. J. Peraire, M. Vahdati, K. Morgan, and O.C. Zienkiewicz. Adaptive remeshing for compressible flow computations. *J. Comp. Phys*, 72:449–466, 1987.

97. S. Prudhomme and J.T. Oden. On goal-oriented error estimation for elliptic problems: Application to the control of pointwise errors. *Comput. Methods Appl. Mech. Engrg.*, 176(1-4):313–331, 1999.

98. R. Rannacher and F.T. Suttmeier. A feed-back approach to error control in finite element methods: Application to linear elasticity. *Comput. Mech.*, 19(5):434–446, 1997.

99. R. Rannacher and F.T. Suttmeier. A posteriori error estimation and mesh adaptation for finite element models in elasto-plasticity. *Comput. Methods Appl. Mech. Engrg.*, 176(1-4):333–361, 1999.

100. U. Rüde. *Mathematical and Computational Techniques for Multilevel Adaptive Methods*, volume 13 of *Frontiers in Applied Mathematics*. SIAM, 1993.

101. J. Sarrate, J. Peraire, and A. Patera. A posteriori finite element error bounds for non-linear outputs of the Helmholtz equation. *Internat. J. Numer. Methods Fluids*, 31(1):17–36, 1999.

102. L.R. Scott and S. Zhang. Finite element interpolation of non-smooth functions satisfying boundary conditions. *Math. Comp.*, 54:483–493, 1992.

103. E.M. Stein. *Singular Integrals and Differentiability Properties of Functions*. Princeton University Press, 1970.

104. H.J. Stetter. *Analysis of Discretization Methods for Ordinary Differential Equations*, volume 23 of *Springer Tracts in Natural Philosophy*. Springer, 1973.

105. J. Stoer and R. Bulirsch. *Introduction to Numerical Analysis*, volume 12 of *Texts in Applied Mathematics*. Springer-Verlag, 1993.

106. B. A. Szabo. Estimation and control of error based on p-convergence. In *Accuracy Estimates and Adaptive Refinements in Finite Element Computations* [32], pages 61–70.

107. P.S. Vassilevski. On two ways of stabilizing the hierarchical basis multilevel methods. *SIAM Rev.*, 39(1):18–53, 1997.

108. P.S. Vassilevski and M.H. Etova. Computation of constants in the strengthened Cauchy inequality for elliptic bilinear-forms with anisotropy. *SIAM J. Sci. Comput.*, 13(3):655–665, 1992.

109. R. Verfürth. A posteriori error estimators for the Stokes equations. *Numer. Math.*, 55:309–325, 1989.

110. R. Verfürth. A posteriori error estimates for nonlinear problems. Finite element discretizations of elliptic equations. *Math. Comp.*, 62:445–475, 1994.

111. R. Verfürth. A posteriori error estimation and adaptive mesh refinement techniques. *J. Comput. Appl. Math.*, 50:67–83, 1994.

112. R. Verfürth. *A review of a posteriori error estimation and adaptive mesh refinement techniques.* Wiley-Teubner, 1996.

113. R. Verfürth. Error estimates for some quasi-interpolation operators. *RAIRO Modél. Math. Anal. Numér.*, 33(4):695–793, 1999.

114. L.B. Wahlbin. Local behavior in finite element methods. In P.G. Ciarlet and J.L. Lions, editors, *Finite Element Methods (Part 1)*, volume 2 of *Handbook of Numerical Analysis*, pages 355–522, North-Holland, 1991. Elsevier Science Publishers B.V.

115. L.B. Wahlbin. *Superconvergence in Galerkin Finite Element Methods*, volume 1605 of *Lecture Notes in Mathematics.* Springer-Verlag, 1995.

116. S. Wang, I.H. Sloan, and D.W. Kelly. Computable error bounds for pointwise derivatives of a neumann problem. *IMA J. Numer. Anal.*, 18(2):251–271, 1998.

117. H. Yserentant. On the multilevel splitting of finite element spaces. *Numer. Math.*, 49:379–412, 1986.

118. E. Zeidler. *Non-linear Functional Analysis and Its Applications II/B.* Springer-Verlag, 1990.

119. Z. Zhang and J.Z. Zhu. Analysis of the superconvergent patch recovery technique and a posteriori error estimator in the finite element method. Part 1. *Comput. Methods Appl. Mech. Engrg.*, 123:173–187, 1995.

120. Z. Zhang and J.Z. Zhu. Analysis of the superconvergent patch recovery technique and a posteriori error estimator in the finite element method. Part 2. *Comput. Methods Appl. Mech. Engrg.*, 163:159–170, 1998.

121. O.C. Zienkiewicz and J.Z. Zhu. A simple error estimator and adaptive procedure for practical engineering analysis. *Internat. J. Numer. Methods Engrg.*, 24:337–357, 1987.

122. O.C. Zienkiewicz and J.Z. Zhu. The superconvergent patch recovery and a posteriori error estimates. Part 1: The recovery technique. *Internat. J. Numer. Methods Engrg.*, 33:1331–1364, 1992.

123. O.C. Zienkiewicz and J.Z. Zhu. The superconvergent patch recovery and a posteriori error estimates. Part 2: Error estimates and adaptivity. *Internat. J. Numer. Methods Engrg.*, 33:1365–1382, 1992.

124. M. Zlamal. Some superconvergence results in the finite element method. In A. Dold and B. Eckmann, editors, *Mathematical Aspects of Finite Element Methods*, number 606 in Springer Lecture Notes in Mathematics, 1975.

125. M. Zlamal. Superconvergence and reduced integration in the finite element method. *Math. Comp.*, 32:663–685, 1978.

Index

PURE AND APPLIED MATHEMATICS

A Wiley-Interscience Series of Texts, Monographs, and Tracts

Founded by RICHARD COURANT
Editors Emeriti: PETER HILTON and HARRY HOCHSTADT, JOHN TOLAND
Editors: MYRON B. ALLEN III, DAVID A. COX, PETER LAX

ADÁMEK, HERRLICH, and STRECKER—Abstract and Concrete Catetories
ADAMOWICZ and ZBIERSKI—Logic of Mathematics
AINSWORTH and ODEN—A Posteriori Error Estimation in Finite Element Analysis
AKIVIS and GOLDBERG—Conformal Differential Geometry and Its Generalizations
ALLEN and ISAACSON—Numerical Analysis for Applied Science
*ARTIN—Geometric Algebra
AUBIN—Applied Functional Analysis, Second Edition
AZIZOV and IOKHVIDOV—Linear Operators in Spaces with an Indefinite Metric
BERG—The Fourier-Analytic Proof of Quadratic Reciprocity
BERMAN, NEUMANN, and STERN—Nonnegative Matrices in Dynamic Systems
BOYARINTSEV—Methods of Solving Singular Systems of Ordinary Differential Equations
BURK—Lebesgue Measure and Integration: An Introduction
*CARTER—Finite Groups of Lie Type
CASTILLO, COBO, JUBETE and PRUNEDA—Orthogonal Sets and Polar Methods in Linear Algebra: Applications to Matrix Calculations, Systems of Equations, Inequalities, and Linear Programming
CHATELIN—Eigenvalues of Matrices
CLARK—Mathematical Bioeconomics: The Optimal Management of Renewable Resources, Second Edition
COX—Primes of the Form $x^2 + ny^2$: Fermat, Class Field Theory, and Complex Multiplication
*CURTIS and REINER—Representation Theory of Finite Groups and Associative Algebras
*CURTIS and REINER—Methods of Representation Theory: With Applications to Finite Groups and Orders, Volume I
CURTIS and REINER—Methods of Representation Theory: With Applications to Finite Groups and Orders, Volume II
DINCULEANU—Vector Integration and Stochastic Integration in Banach Spaces
*DUNFORD and SCHWARTZ—Linear Operators
Part 1—General Theory
Part 2—Spectral Theory, Self Adjoint Operators in Hilbert Space
Part 3—Spectral Operators
FARINA and RINALDI—Positive Linear Systems: Theory and Applications
FOLLAND—Real Analysis: Modern Techniques and Their Applications
FRÖLICHER and KRIEGL—Linear Spaces and Differentiation Theory
GARDINER—Teichmüller Theory and Quadratic Differentials
GREENE and KRANTZ—Function Theory of One Complex Variable
*GRIFFITHS and HARRIS—Principles of Algebraic Geometry
GRILLET—Algebra
GROVE—Groups and Characters
GUSTAFSSON, KREISS and OLIGER—Time Dependent Problems and Difference Methods

HANNA and ROWLAND—Fourier Series, Transforms, and Boundary Value Problems, Second Edition
*HENRICI—Applied and Computational Complex Analysis
Volume 1, Power Series—Integration—Conformal Mapping—Location of Zeros
Volume 2, Special Functions—Integral Transforms—Asymptotics—Continued Fractions
Volume 3, Discrete Fourier Analysis, Cauchy Integrals, Construction of Conformal Maps, Univalent Functions
*HILTON and WU—A Course in Modern Algebra
*HOCHSTADT—Integral Equations
JOST—Two-Dimensional Geometric Variational Procedures
*KOBAYASHI and NOMIZU—Foundations of Differential Geometry, Volume I
*KOBAYASHI and NOMIZU—Foundations of Differential Geometry, Volume II
LAX—Linear Algebra
LOGAN—An Introduction to Nonlinear Partial Differential Equations
McCONNELL and ROBSON—Noncommutative Noetherian Rings
NAYFEH—Perturbation Methods
NAYFEH and MOOK—Nonlinear Oscillations
PANDEY—The Hilbert Transform of Schwartz Distributions and Applications
PETKOV—Geometry of Reflecting Rays and Inverse Spectral Problems
*PRENTER—Splines and Variational Methods
RAO—Measure Theory and Integration
RASSIAS and SIMSA—Finite Sums Decompositions in Mathematical Analysis
RENELT—Elliptic Systems and Quasiconformal Mappings
RIVLIN—Chebyshev Polynomials: From Approximation Theory to Algebra and Number Theory, Second Edition
ROCKAFELLAR—Network Flows and Monotropic Optimization
ROITMAN—Introduction to Modern Set Theory
*RUDIN—Fourier Analysis on Groups
SENDOV—The Averaged Moduli of Smoothness: Applications in Numerical Methods and Approximations
SENDOV and POPOV—The Averaged Moduli of Smoothness
*SIEGEL—Topics in Complex Function Theory
Volume l—Elliptic Functions and Uniformization Theory
Volume 2—Automorphic Functions and Abelian Integrals
Volume 3—Abelian Functions and Modular Functions of Several Variables
SMITH and ROMANOWSKA—Post-Modern Algebra
STAKGOLD—Green's Functions and Boundary Value Problems, Second Editon
*STOKER—Differential Geometry
*STOKER—Nonlinear Vibrations in Mechanical and Electrical Systems
*STOKER—Water Waves: The Mathematical Theory with Applications
WESSELING—An Introduction to Multigrid Methods
†WHITHAM—Linear and Nonlinear Waves
†ZAUDERER—Partial Differential Equations of Applied Mathematics, Second Edition

*Now available in a lower priced paperback edition in the Wiley Classics Library.
†Now available in paperback.